The TAB Electronics Technician's On-Line Resource Reference

Stephen J. Bigelow

McGraw-Hill

New York   San Francisco   Washington, D.C.   Auckland   Bogotá
Caracas   Lisbon   London   Madrid   Mexico City   Milan
Montreal   New Delhi   San Juan   Singapore
Sydney   Tokyo   Toronto

Library of Congress Cataloging-in-Publication Data
Bigelow, Stephen J.
    The TAB electronics technician's on-line resource reference /
Stephen J. Bigelow.
        p.   cm.
    ISBN 0-07-036219-X (trade). — ISBN 0-07-036220-3 (paper)
    1. Electronic apparatus and appliances—Maintenance and repair—
Computer network resources—Directories.   2. Microcomputers—
Maintenance and repair—Computer network resources—Directories.
3. Online data processing—Handbooks, manuals, etc.   I. Title.
TK7805.B54   1996
025.06'621381—dc20   96-42140
                                                                CIP

McGraw-Hill

A Division of The McGraw·Hill Companies

1 2 3 4 5 6 7 8 9 0   DOC/DOC   9 0 1 0 9 8 7 6

ISBN 0-07-036220-3 (PBK)
ISBN 0-07-036219-X (HC)

*The sponsoring editor for this book was Steve Chapman, the editing supervisor was
Sally Glover, and the production supervisor was Pamela Pelton. It was set in Galliard
by North Market Street Graphics.*

Printed and bound by R. R. Donnelley & Sons Company.

 This book is printed on recycled, acid-free paper containing a
minimum of 50% recycled, de-inked fiber.

McGraw-Hill books are available at special quantity discounts to use as premiums and
sales promotions, or for use in corporate training programs. For more information, please
write to the Director of Special Sales, McGraw-Hill, 11 West 19th Street, New York, NY
10011. Or contact your local bookstore.

This book is dedicated to every computer enthusiast, technician, and engineer who ever needed a product specification, jumper setting, or software patch.

Contents

Disclaimer and Cautions

It is important that you read and understand the following information. Please read it carefully!

PERSONAL RISK
AND LIMITS OF LIABILITY

Neither the author, the publisher, nor anyone directly or indirectly connected with the publication of this book shall make any warranty either expressed or implied, with regard to this material, including, but not limited to, the implied warranties of quality, merchantability, and fitness for any particular purpose. Further, neither the author, publisher, nor anyone directly or indirectly connected with the publication of this book shall be liable for errors or omissions contained herein, or for incidental or consequential damages, injuries, or financial or material losses resulting from the use, or inability to use, the material contained herein. The material is provided AS-IS, and the reader bears all responsibilities and risks connected with its use.

REFERENCE WARNING

The manufacturers, service providers, distributors, and other contacts listed and presented in this book are shown for reference purposes *only*. Their mention and use in this book shall not be construed as an endorsement of any individual or organization, nor the quality of their products or services, nor their performance or business integrity. The author, publisher, and anyone directly or indirectly associated with the production of this book expressly disclaim all liability whatsoever for any financial or material losses or incidental or consequential damages that might occur from contacting or doing business with any such organizations or individuals.

The information, data, comments, descriptions, software, and all other materials in any form available through on-line resources remain the exclusive property of their respective copyright holder(s). The author, publisher, and anyone directly or indirectly associated with the production of this book expressly disclaim all liability whatsoever for the accuracy or content of any on-line resource. The opinions, views, and outright attitude found on-line are not necessarily those of the author or publisher.

VIRUS WARNING

There is a proliferation of software available through on-line sources. However, there is also a possibility that the software you obtain might be infected with a computer virus. You are strongly cautioned to use a current virus checking utility to examine any software you obtain before executing that software for the first time. The author, publisher, and anyone directly or indirectly associated with the production of this book expressly disclaim all liability whatsoever for any financial or material losses or incidental or consequential damages that might occur from a computer virus or other rogue software.

Introduction

It really doesn't matter whether you're an experienced technician or if you are installing your first SIMM, the successful upgrade and repair of any piece of electronics depends on three things: the right tools, the right parts, and the right information. Without these three vital elements, your chances of performing efficient, cost-effective service are truly diminished. Of these three elements, the right *information* is certainly the most elusive. Who has the parts you need? How do you install and configure them? If there's a problem, how do you get tech support? Sure, you can call the manufacturer directly, but live technical support is rapidly becoming a thing of the past. Try a tech-support call and you'll see what I mean. Just when you think you've been on hold forever, a "technician" comes on and doesn't have an answer (or worse, "guesses" at an answer), which wastes even more of your time. Now don't get me wrong. The high-tech industry has its share of bright, hard-working support people, but in today's world of shrinking support budgets and short product life cycles, finding answers to your questions is harder than ever.

As companies attempt to automate their support, they are making use of on-line resources, such as private bulletin board services (BBS) and fax-back systems, to provide "how-to-do-it" information. Many companies have established forums on commercial services such as CompuServe, America Online, and Prodigy. Even more recently, the Internet has become extremely popular for its worldwide e-mail, file transfer, and World Wide Web capabilities. And on-line information is going far beyond companies. Universities, individuals, user groups, and a myriad of "after-market" organizations have established themselves on-line. Their varied experience and research have brought a wealth of electronic and computer information to services such as the Internet. The problem with on-line resources is *finding* them in the first place.

This book is intended to cut through the confusion and mystery of the growing on-line resources by providing a comprehensive, well-researched index of on-line sources. If you have a PC and a modem, you can typically find answers to your technical questions on-line, saving long-distance toll charges, lots of time, and plenty of frustration in the process.

WHAT THE BOOK COVERS

For anyone just starting out on-line, Chapter 1 provides an introduction to the on-line services that are available to you, especially the Internet. Chapter 2 takes that basic introduction a few steps further by offering tips and techniques that will help you get the most from on-line services. The real heart of the book is Chapter 3. With over 1200 industry contacts (over 100 of which are illustrated), this is the first on-line reference dedicated exclusively to electronics and computer resources. In addition to the company references in Chapter 3, Chapter 4 provides an extensive index of electronics and computer-related news groups and mailing lists, each focused on a particular topic of interest. If the manufacturer can't answer your questions, the resources in Chapter 4 put you in touch with the combined resources of thousands of experienced users and technical professionals. Chapter 5 is a cross-reference that allows you to locate the manufacturers of Chapter 3 by topic or specialty area. This allows you to find an array of related manufacturers quickly. An on-line glossary puts an index of typical on-line terms at your fingertips. Appendices offer a list of international Internet providers, as well as a wealth of other on-line-related information.

BUT YOU'VE MISSED MY SITE!

Like any reference book, *The TAB Electronics Technician's On-line Resource Reference* should be considered a "work in progress;" there are undoubtedly BBS numbers, fax-back services, web sites, and even entire companies that simply did not make it into this edition. That is why we have established "*The COM Port*," a regular column to our premier troubleshooting newsletter, *The PC Toolbox*™. In addition to regular PC and peripheral troubleshooting articles, we will keep you up to date with new on-line resources as they become available. If you know a site we've missed (or have an electronics or PC site of your own), tell us about it so we can get it into our next issue/edition.

WE'RE INTERESTED
IN YOUR SUCCESS

We want this book to be a reliable resource that you can turn to on a regular basis. Tell us what we can do to make this book better!

Dynamic Learning Systems
Attn: Stephen J. Bigelow
P.O. Box 282
Jefferson, MA 01522-0282 USA
Tel: 508-829-6744
Fax: 508-829-6819
TechNet BBS: 508-829-6706 (up to 28.8KB 8 data bits, 1 stop bit, no parity bit)
CompuServe: 73652,3205
Internet e-mail: sbigelow@cerfnet.com
WWW: http://www.dispubs.com/

CHAPTER

1

An On-line Primer
for Technicians

There is little doubt that the *Internet* has become an icon of the information age. The endeavor that began as a government networking experiment in the late 1960s has quickly evolved into a complex series of networks that literally extends around the world. With thousands of university, government, and private industry networks already joined together, the Internet represents an unprecedented information resource for individual computer users and businesses of all sizes. According to *The Internet Society*, there are already 30 million or more active Internet users—a number that is swelling by 160,000 users each month. This chapter illustrates the essential concepts of the Internet and other on-line services, shows you what is involved in going on-line, and explains the scope of on-line services that is available.

UNDERSTANDING THE INTERNET

The first problem in understanding the Internet is to really know what the Internet *IS*. Expressed simply, the Internet is nothing more than a "network of networks." While this is a stunning oversimplification, it is basically what the Internet is all about. You see, although individual networks are owned and operated by their particular university or research center, no one really "owns" the Internet; there is no centralized authority or administration as there is with other commercial on-line services, such as CompuServe or America Online. Even though thousands of networks are linked together to form the Internet, each network is totally independent. This is largely the reason why the Internet is so unregulated (and why it has taken so long for individual users and ordinary businesses to have easy access). This is also the reason why the Internet can be so overwhelming to new users; even though you can now access thousands of networks, it does you no good at all unless you know *what* you are looking for and *where* it is located. It's rather like being afloat on a sea of information without a compass.

A Bit of History

To really appreciate what the Internet has to offer, you should know a bit about the evolutionary process that shaped it. In the early 1970s, the US Defense Department wanted to build a computer network that would remain functional in the event of a nuclear war. This prototype network was called ARPAnet. Because reliability was essential, communication was intended to take place between two computers directly, any portion of the network could be disabled, and communication would assume one of several alternative paths to the destination (an early example of *peer-to-peer networking*). The message being sent was enclosed in a standardized "envelope" known as an *Internet Protocol* (IP) packet. The IP packet contained the "address" of the destination machine.

The ARPAnet project continued into the early 1980s along with the development of local-area networks (LANs). Most LAN workstations in the early 1980s ran the UNIX operating system, which came with IP networking built-in. Organizations with their own LANs preferred to connect to the ARPAnet rather than a single time-sharing computer at their own location. Because UNIX supported IP communication, a connection proved rather straightforward, although using the system on the ARPAnet required proficiency with UNIX.

By the late 1980s, the National Science Foundation (NSF) built their own network (called NSFNET), which was based on ARPAnet's IP communication technology. NSFNET was designed to support five supercomputer centers located around the country. However, each center needed to be connected together, as well as provide access to researchers and academic clients. Because dedicated telephone lines were expensive, the NSF created regional networks, which allowed each site to be connected in a daisy-chain fashion, and it allowed connections for local clients. Thus, messages could be exchanged from computer to computer by passing messages back and forth along the daisy chain. This architecture was the forerunner of what we know as the Internet.

By the late 1980s, Merit Network, Inc. upgraded the NSF network and opened access to academic researchers, government organizations, and contractors. Access was also extended to other countries allied to the United States. Today, the network provided by the NSF forms the backbone of the Internet. But that backbone is joined by over 5000 regional, state, campus, and corporate networks. Links to Canada, Europe, Japan, Australia, Central America, and South America are now operational.

The Role of Commercialization

Like all government works, use of the Internet for "personal gain" has largely been prohibited. However, commercial organizations see the Internet as a very viable outlet for selling goods and services. By 1992, restrictions on the commercial use of the Internet were revoked. Today, there are actually more commercial sites using the Internet than educational and research organizations. That trend will almost certainly continue as more users and businesses come on-line.

WHY USE THE INTERNET?

The size and scope of the Internet is truly impressive, but most ordinary individuals stop at this point and ask the next obvious question: "What good is all of this to *me*?" True, Internet resources are staggering, and unless you have a specific need, there will be a great deal of information that you will probably never use (at least initially). However, there are four compelling activities that any first-time Internet user can benefit from: e-mail, network news, file transfers, and information browsing.

E-mail

Electronic mail (or e-mail) is just as the name implies. You can send and receive messages with anyone on the Internet, as well as anyone using on-line services with Internet gateways (such as CompuServe, America Online, Delphi, and so on). E-mail is typically much faster, and generally much cheaper, than traditional "post office" mail. Many manufacturers are starting to use the Internet as the forum of choice for national and international tech-support. As a result, your product questions and problems can typically be answered faster and with less "time-on-hold" than a telephone call. There is little doubt that "live" tech-support is fading fast as manufacturers try to reduce their load of clogged telephone lines

News Groups and Mailing Lists

There is an incredible variety of electronic discussion groups (or *network news groups*) available on the Internet. Topics range from the mundane to the bizarre. By "subscribing" to each news group of interest (virtually all news groups are free), you can stay in touch with the very latest discussions, opinions, and happenings. An alternative to the news group is the *mailing list,* which echoes copies of pertinent e-mail to all members of the particular list(s). This allows you to keep abreast of the latest news and information in areas that interest you. Like news groups, there are a wide variety of mailing lists available on the Internet.

File Transfers

There is a multitude of files and utilities available on the Internet in the form of shareware and freeware. Many manufacturers post new drivers and program patches on their FTP sites. Using file transfer protocol (FTP) utilities, you can retrieve or upload files from a computer in Tokyo as easily as you could from a computer in the next room or down the street; it really doesn't matter where the other computer is located.

Information Browsing

As an Internet user, you can find everything from weather reports and space photographs to government databases located on features such as Gopher or the World Wide Web. Traditionally, the problem is knowing what's available and knowing where to find it. There are few guides to resources because there is simply too much change and growth. Fortunately, there are a growing number of utilities that make it possible to navigate easily through Internet resources in order to find what *is* available. Browsing tools allow you to traverse from system to system and review their offerings in a matter of moments.

INTERNET CONNECTIONS

Connecting to the Internet has always been somewhat of a chore. This was due in part to the lack of central authority, and in part to the lack of personal computers powerful enough to interact with the Internet effectively. Only a few years ago, getting an Internet account was a matter of who you knew. It typically required a mainframe and network at a university or research organization before an Internet connection could be established. Then, individual accounts were distributed as needed to students, educators, researchers, or whoever had a good reason to be on-line. Even as PCs and operating systems became powerful enough to "stand alone" on Internet, the problem was *where* to connect. This later problem has largely been resolved by the explosive growth of local, national, and global Internet service providers who have their networks connected to the Internet, and sell access on their networks to individuals, businesses, and other networks.

The Service Provider

You can't simply splice a cable and wind up with Internet access; there is much more involved. But regardless of how you choose to go on-line, an Internet service provider will be your most valuable friend and ally. A comprehensive list of global Internet service providers is presented in Appendix A. Remember that the Internet is not centrally managed; there is no one place you can go to sign up. A service provider is a company that owns and operates a network that has access to the Internet and, in turn, the service provider can supply you with several types of connections. The service provider charges you a fee for your connection (telephone toll charges are separate), and a portion of those proceeds go to support the Internet as a whole. If you want to get a current listing of available service providers for your own region or country, try visiting the search engine on the World Wide Web at: http://thelist.com/.

It is important for you to remember that not all service providers are created equal. The quality, cost, and reliability of their services can vary radically between providers. Some service providers offer regional coverage, while others offer national (or even global) support. Because each provider is owned and managed independently, they might not suit your needs—and some providers are simply better than others. Your best course is to shop around and compare the deals offered by each provider. Today, a service provider can offer three typical paths to the Internet: the direct connection, the dial-in connection, and the on-line service. Each path carries its share of requirements and costs, so consider your needs carefully before proceeding.

Direct Connection

A direct connection is clearly the most powerful and expensive form of Internet connection. In effect, you must have (or be part of) a network. Then you need a dedicated communication line (such as a T1 56Kbps line or a T3 line), along with hardware and software (typically purchased or leased from your service provider) that supports your connection to the service provider.

A direct connection offers some distinct advantages. You can set up an FTP server that allows outside individuals and organizations to access your systems; in effect, you become a tangible new part of the Internet. Commercial organizations moving to do business on the Internet often take advantage of that capability. Next, a direct connection keeps you permanently on-line. And third, at 56Kbps or more, file transfers are accomplished much faster.

Unfortunately, the direct connection carries a substantial cost. A dedicated connection can easily cost an organization $15,000 to $20,000 or more in the first year, and a bit less each subsequent year. However, you can see that a direct connection is reserved for organizations that need them and can afford them.

Dial-In Connection

As an alternative to direct connections, most service providers offer dial-in (or *dial-up*) connections. The dial-in connection is a medium-performance connection that requires only a PC, Internet TCP/IP communication software, and a 14.4KB or faster modem. A 9600 bps modem often works, but access can be painfully slow. The dial-in account works much like other standard on-line services; you dial a number to your service provider, establish a modem connection, and enjoy your Internet access. The PC itself should use an i386 or faster CPU, 4MB of RAM or more, DOS 5.0 or later, Windows 3.1 or later, and at least 10MB of hard drive space.

The dial-in connection provides virtually all of the services that a direct connection would (except the ability to implement your own FTP or web server), but the connection is only active while the modem connection is established. Thus, the cost is only a per-hour connect fee (usually under $100 for 20 hours per month of service). If you are making a toll call for access, you can expect to pay a toll fee as well (although some national service providers supply 800 or local-access numbers). Even with toll charges, a dial-in connection is the preferred type of Internet access for individuals and small businesses.

There are two types of dial-in connections: SLIP and PPP. As an Internet user, you do not have to worry about the technical intricacies of SLIP and PPP; you simply have to ensure that the software and TCP/IP drivers used on your PC match the connection type. Most service providers can easily support SLIP accounts, and much of the bundled software out there is written for SLIP operation. PPP is a slightly newer and more effective protocol implemented in Windows 95 Internet services, but fewer service providers support it, and less of the available software is designed to take advantage of it.

Dial-in connections are greatly simplified by the trend toward *bundling* or packaging Internet software and TCP/IP drivers with books. There are a number of good starter bundles out there, but I got on-line with the *Internet Membership Kit* from Ventana Press. Although its *Chameleon Sampler* software was not the simplest software to configure, it had everything I needed to go on-line. Take a walk through your local PC store and you'll probably find a whole variety of ready-to-go dial-in packages. Once you are finally on-line, you can download (or "FTP") additional shareware or freeware utilities that offer even better mail, news, or browsing capabilities. Regardless of the bundle that you buy, the process is generally the same:

- Install the bundled software under Windows.
- Choose a service provider—publishers that bundle books and software generally strike *excellent* deals with national service providers.
- Call the chosen provider by phone or BBS to establish the account.
- Wait a day or two for the account to be activated.
- Configure the bundled software and start your access.

On-line Services

Of course, there are alternatives to your own Internet access. An on-line service such as CompuServe, America Online, Delphi, and Prodigy already provide Internet access as part of the regular service fee. If you subscribe to an on-line service or high-end bulletin board service (BBS), you might also have access to the Internet—the number and quality of the services that are offered depend on the particular facility. The advantage of an on-line service is *simplicity;* you can browse extensive archives in a wide proliferation of forums, exchange e-mail, and access the Internet, all through a single service with highly integrated software. Costs are generally quite low, while content and organization are typically high. If you already use a commercial service (or the thought of finding an Internet provider scares you), chances are that you already have access to the Internet. There is more about commercial services later in this chapter.

ADDRESSING AND NAMING CONVENTIONS

Of all the subtle complexities of the Internet, few evoke more frustration and confusion among new users than *naming conventions.* On the surface, that collection of numbers and nonsense words might seem arcane—even random. But if you look closer, you will see a method to the madness. This part of the chapter is intended to demystify Internet addressing and naming conventions.

Addressing

Each and every computer connected to the Internet has two unique addresses: an IP address and a domain name. The *IP address* is a series of four numbers (each less than 256) that are separated by periods. For example, a typical IP address might look like 134.84.101.1. For the most part, you do not have to concern yourself with remembering IP addresses—just the domain name.

Naming

There are actually several types of names involved in the naming process: the user name, the domain name, and the top domain name. The *user name* is the unique alphanumeric string that identifies you (actually your computer) as an individual. A user name can be completely arcane (such as **axc13ex**), but if you are in a position to choose your own user name, pick one that will be meaningful to you and the people who will be sending you e-mail. For example, my user name is **sbigelow**. The *domain name* generally refers to the supporting network (or service provider) that your computer is connected to, and is separated into two parts: the domain name itself, followed by a three-character *top domain* name. For my account, the domain name would be **cerfnet**, and the top domain would be **com**. Taken together, my Internet name is **sbigelow@cerfnet.com**—pronounced "s bigelow at surf-net dot com." Feel free to send me some e-mail when you get on-line.

Of course, the names we have talked about so far are pretty simple, but what happens with an address like, **j_jones@rabbit.oit.unc.edu**? Well, reading the address gets a bit more complicated, but the basic rules are the same. First, the user name (j_jones) is easy enough to find. The term right after the @ (pronounced "at") symbol is the computer's name (rabbit), and the top domain name (edu) identifies the address as an educational institution. The symbol (oit) represents *Office of Information and Technology,* while the symbol (unc) indicates the *University of North Carolina.* Unfortunately, you are hard-pressed to know that simply by looking at the symbols. What this means for you is that not all addresses are self-explanatory, but there is a practical reason for everything you see.

You can tell a bit about the type of domain from looking at the top domain name. The three-character code explains what type of organization the domain represents. Table 1-1 shows six top domain names that are generally used in the United States. For users outside of the US, the top domain name is a two character code as shown in Table 1-2.

TOOLS OF THE TRADE

Once you are finally on-line with a dedicated or dial-in connection, you will be faced with the challenges of exploring the Internet. Over the years, a variety of tools have developed to serve each individual function, and each small

TABLE 1-1 Typical top domain names used in the United States

com	A commercial business or organization
edu	An educational institution such as a college or university
gov	A government department or organization
mil	A military installation or organization
net	Network resources—an integral part of the Internet infrastructure
org	Typically a nonprofit or other organization

application (or "applet") works together to form a set of utilities (rather like the generic utilities under the Windows Application icon). This part of the chapter helps you match the tools to the tasks. For the purposes of this discussion, let's assume that you're an individual user who has purchased bundled software and documentation.

The Basic Internet Toolkit

The tools that come with your bundled software are certainly useful and functional, but few packages contain the full-blown versions of any software tools. This means that while you'll be able to perform a variety of functions, you probably can not take full advantage of the Internet. At the very least, a bundled software package will offer five working areas:

- *The TCP/IP Stack* This is a vital background utility that allows your Windows PC to communicate with your service provider through

TABLE 1-2 Typical top domain names used outside of the United States

au	Australia
at	Austria
bz	Belize
ca	Canada
dk	Denmark
fi	Finland
fr	France
de	Germany
it	Italy
jp	Japan
no	Norway

standard TCP/IP. The TCP/IP stack is typically loaded before dialing your service provider, and it unloads automatically after the connection is severed. Without this function, your individual dial-in connection could not work.

- *E-mail Utility*   Virtually all bundled Internet software offers some form of e-mail utility that allows you to retrieve and create your own e-mail. Mail utilities vary in quality and efficiency, but they get the job done.

- *FTP Utility*   Part of the power of the Internet is your ability to get into other computers around the world to upload or download files. Software bundles typically provide a basic FTP utility. Using FTP, you can download articles, electronic books, and other more powerful Internet utilities.

- *Telnet Utility*   While an FTP function merely gives you access to another computer's directories and files, Telnet actually allows you to work with other computers in an interactive fashion through a command-line interface. You can still upload and download files.

Power Tools for the Internet

While bundled software can get you onto the Internet, they can only take you so far. Fortunately, there are many more shareware and freeware programs that make it easier for you to take advantage of Internet resources. The following list of programs are all available right off the Internet. Their locations are listed for your convenience.

- *Eudora*   This is one of the best Internet all-purpose e-mail programs available. If you find yourself handling a great deal of e-mail, consider downloading the freeware version of Eudora by FTP from ftp. qualcomm.com in the /pceudora/windows/ directory.

- *Gopher*   Tools such as e-mail and FTP utilities are great for specific tasks, but the vast majority of Internet resources are buried under layers of networks and directories. Moving from one place to another can be a long, arduous task. What is needed is a "browser" that lets you move from one place to another as efficiently as possible. The Gopher tool is perhaps the most effective browser for the Internet. Although I cringe to say it, Gopher is a tool that lets you "surf the net." You can find a freeware copy of Gopher by FTP from lister.cc.ic.ac.uk in the /pub/wingopher/ directory.

- *Trumpet*   Another feature of the Internet is the availability of network news groups. Basically, network news groups are electronic dis-

cussion areas where people can post questions and exchange information. To subscribe to and participate in news groups, you will need a
news reader utility such as Trumpet. Trumpet is available as shareware, and can be downloaded by FTP from ftp.utas.edu.au in the
/pc/trumpet/wintrump/ directory.

- *Mosaic/Netscape* One of the most exciting developments on the
  Internet is the World Wide Web (WWW), which offers a vast array of
  documents combining text, images, sound, and video. Mosaic is a tool
  designed to browse the WWW; it also provides limited FTP, news, and
  Gopher services. Mosaic is freeware available by FTP from ftp.
  ncsa.uluc.edu. A more recent (and more popular) web browser is the
  Netscape Navigator 2.0 available from ftp.netscape.com.

UNDERSTANDING OTHER RESOURCES

As vast as it is, the Internet is not the only on-line source available to technicians. Commercial on-line services, faxback services, and a host of private BBS
facilities offer a wealth of information for the casual PC user and hard-core
technician alike. This part of the chapter explains each of these varied (but no
less important) alternatives.

Technical Support Telephone

The use of automation in telecommunication has exploded over the past few
years. Today, it is almost impossible to call a major corporation without
encountering an automated answering system (referred to as *voice mail*). The
same technology that makes voice mail possible has also provided automated
support. Typically, 10 percent of the technical questions received by a manufacturer are encountered 90 percent of the time, so by listing those questions
and their answers in an automated menu, a technician can call in and search
for answers via the telephone without any human intervention at all on the
manufacturer's end. While this might not really qualify as an "on-line
resource" in the conventional sense, it is often a worthwhile place to start a
search for answers. However, only the very largest companies have the per-
sonnel and expertise to implement and maintain such automated systems.

Fax and FaxBack

The facsimile (or *fax*) has proven to be one of the most important tools available for business communication. Unfortunately, it is also one of the least effi-

cient for exchanging correspondence. Writing takes a lot of time—for the person asking *and* for the person answering. Ultimately, you can send a tech support question by fax, but don't expect an answer anytime soon.

By contrast, faxback services have become an important resource for both enthusiasts and technicians alike. Where voice mail can provide automated vocal answers to common questions, faxback services offer automated document distribution. By calling the service and selecting from a menu of available documents, you can get documents faxed back to your own fax machine in a matter of minutes. For example, you can call a hard drive maker's faxback service and request installation instructions, jumper settings, and drive geometries for a particular drive. Because there is no human intervention on the other end, the response time is usually under ten minutes. When you find yourself faced with a configuration issue (and no manual is at hand), a faxback is usually your best bet for rapid resolution of the problem.

Bulletin Boards

Bulletin boards have been popular since the late 1980s when modems became fast enough to exchange text and image data efficiently. Simply stated, a BBS is little more than a stand-alone PC and a modem running BBS software. The BBS software allows files (i.e., shareware programs, file patches, and text documents) to be made available to users that can call in from any remote computer. For a technician, the BBS offers access to documents, shareware, drivers, patches, and so on. It is also possible to exchange messages over the BBS. Taken together, the BBS is a versatile, automated resource that is easy to use, relatively inexpensive to assemble, and simple to maintain. Chapter 3 lists manufacturer's BBS numbers wherever they are available.

There are some limitations to the BBS. First, there is usually a toll charge involved in the actual call (unless the BBS is providing an 800 number). Second, the number of users that can use a BBS at any one time is limited. Third, the time on-line is often limited (due to the fact that there are only a finite number of users that can use the BBS at any one time) to anywhere from 15 to 60 minutes.

CompuServe

The CompuServe Information Service (called *CIS* and based in Columbus, Ohio) has been one of the more popular commercial services available. With a broad mix of manufacturer's forums, special-interest forums, and file archives, CIS proved to be a major on-line resource for technicians before the Internet opened up to commercial use. Excellent content and discussion groups have attracted a large number of companies and manufacturers, so expect to find lots of CIS-based technical support. CIS services are available world-wide, have a reputation for reliability, and offer a very competitive rate schedule. As a commercial service, CIS did not originally provide indirect access to the Internet, but today, its services include FTP, Telnet, newsgroups, and web access (developed by Spry which CompuServe acquired). You can obtain more information about CIS by contacting their customer support at 800-848-8199 or http://www.compuserve.com/

America Online

At its core, America Online (or *AOL*, based in Vienna, Virginia) is very much like other top commercial services. It provides an excellent variety of forums and chat groups, as well as extensive software archives. AOL also focuses much more on flashy art and eye-catching icons. While AOL has grown rapidly over the past few years, it does not appear to have the same appeal to manufacturers that CompuServe has. Still, AOL has positioned itself aggressively in the area of Internet services, and sharp-looking web maps guide the user to all of the Internet's major services. For those who are shy about going on-line, AOL is usually a good first choice because there is excellent on-line instruction and help. You can contact AOL customer service at 800-827-6364 or http://www.aol.com/

Delphi

Appealing more to low-end on-line users, Delphi (based in New York City, New York) has lagged a bit behind other services, largely because of a mediocre interface and installation instructions. Though the range of content and discussion groups are generally regarded as good, Delphi has not commanded the same attention of manufacturers that other services have. Mediocre e-mail services have also made it more difficult to establish effective on-line technical support. Still, Delphi has enjoyed a long tradition of Internet access, and as Delphi continues to update its hardware, the Internet service should at least remain satisfactory. Keep in mind that Delphi's web access is text-only, which makes it a slightly more cumbersome interface for beginners. You can contact Delphi customer service at 503-323-1000 or http://www.delphi.com/.

Prodigy

The Prodigy service (based in White Plains, New York) has been designed partly as a user-friendly, inexpensive on-line resource for casual home users, and partly as an outlet for commercial advertising. This is shown by Prodigy's excellent instructions and discussion groups. But without a strong base of manufacturer's forums, technicians generally have limited use for Prodigy alone. Yet Prodigy has added Internet access to its range of services, focusing primarily on web access, e-mail, and newsgroups. Contact Prodigy customer service for more information at 914-448-8000.

2

Using On-line Resources

It is impossible to overstate the importance of technical information. The ability to quickly locate product data, specifications, demo products, software patches and upgrades, and other technical details not only enhances your productivity, but also makes your life easier. Manufacturers, service organizations, and even talented individuals are placing ever-more information on-line. As discussed in Chapter 1, there are a myriad of information resources available to you. Faxes, BBS systems, commercial services, and the Internet are rapidly gaining importance. This chapter explores each of those services in more detail, and provides some helpful guidelines for getting the most from each resource.

MAKING THE MOST
OF FAX-BACK SERVICES

Fax-back systems (also referred to as *automated document distribution systems*) are basically little more than dedicated bulletin board systems running a voice-command menu. You call a fax-back system just as you would any other voice telephone line, but instead of speaking with a person, you hear an introduction and the options of a main menu. By pressing the corresponding key(s) on your TouchTone telephone (rotary telephones will not work), you can make selections to navigate the fax-back system.

Getting the Job Done

In most cases, you need to provide a fax-back system with three pieces of information: the document number(s) that you require, the telephone number of your fax machine (where the documents are to be sent), and your own personal number, which identifies you as the fax recipient. This last feature is used so that if you are calling from a busy business, where one fax machine serves many people, whoever picks up the fax-back document will know it belongs to you. Of course, if you have your own fax machine, that personal number is not important, and you can enter anything you wish. Don't worry about when and how to enter your information; the fax-back system provides you with voice prompts that walk you through every step of the process.

Fax-back systems might hold thousands of documents, and virtually all systems have a single catalog or document list. If you are not sure just which fax you need (or you want to see everything that is available), start by ordering the document list, and have that faxed to you first. You can call the fax-back system again to order the specific document(s) of interest. Keep in mind that some fax-back systems (especially busy systems) limit the number of documents that you can order during any one call. If you want more documents than you can get during a call, you can just make additional calls later.

Keep in Queue

Once documents are ordered, they enter a queue. Depending on just how backed-up the system is, it might take up to 30 minutes before the system attempts to send your documents. If your fax machine happens to be busy at that moment, the fax-back system will usually make several more attempts to send the documents before abandoning the request. For best results, make your requests at times when your fax machine is most likely to be idle. If your fax machine tends to be busy, try and get the documents sent to a quieter machine (e.g., another department).

Check Your Fax Paper

Finally, remember that some fax-back documents can be quite large, especially if you are ordering several of them. As a consequence, make sure that there is plenty of fax paper in your machine before ordering your documents.

NAVIGATING A BBS

By the mid-1980s, computers and modems had become powerful enough to support reasonably fast data communication. This gave rise to a whole new generation of communication software and tools. Among these tools was the Bulletin Board System (or *BBS*). In its simplest form, a BBS is a stand-alone computer and modem running suitable BBS communication software. When configured properly, a BBS can manage the exchange of written messages, as well as the distribution of downloads (such as software patches, demos, and shareware) and perform all of those operations without human intervention.

The relative simplicity, reliability, and low cost of these "automated information managers" made them particularly well suited for technical support. For many companies, the BBS was their first venture into on-line support, and they quickly proved their value by providing service 24 hours per day, 7 days per week. The ability to download software patches and demos allowed companies to greatly expand their presence without the time and expense of mailing diskettes. In fact, the entire shareware industry is built on a distribution network of thousands of independently operated BBS systems across the US and around the world. Although dwarfed and outdated by the Internet, BBS facilities such as TechNet BBS continue to be an important source of shareware and technical information.

Making the Connection

You need four things to connect to a BBS: a PC, a modem, a telephone line, and communications software. There are a wide variety of communication software packages available today (such as SmartCom or ProCom Plus), but if you are new to the on-line world, you might prefer to get your feet wet using "vanilla" communication software such as Terminal (with Windows 3.1) or HyperTerminal (under Windows 95). There are few bells and whistles with these packages, but they come with the operating systems already. Configure your communication software for the proper data frame, most BBS systems use 8 data bits, no parity bit, and 1 stop bit (8/N/1) as shown in Fig. 2-1. Also check to see that the proper AT command strings are entered for your particular modem (more about the AT command set is presented in Appendix C). Once the software is configured properly, you are ready to dial your BBS. If you are new to a BBS, give TechNet BBS a try at 508-829-6706.

Pay the Toll

One of the major disadvantages to a BBS is that the call often demands toll charges. Unlike commercial on-line services and many Internet providers, which offer "local access," calling a BBS is just like calling a person, and the

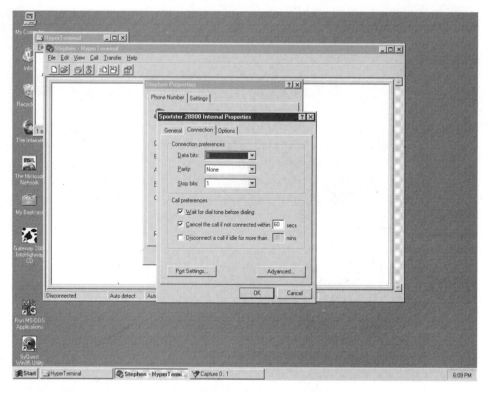

Figure 2-1 Configuring your communication software.

call is subject to the same long-distance and time-of-day rates that ordinary phone calls are. Even if the use of a BBS is free, the phone bill is not, so always remember to keep an eye on the clock when using a long-distance BBS. You can save on toll charges by waiting until evenings or weekends when toll charges are lower. When you are on-line, rather than searching through file areas looking for files, download the "index" files, which can be read off-line in a word processor. For example, TechNet BBS provides the ALLFILES.ZIP and NEWFILES.TXT files. Once you read through these files at your convenience, you will know exactly which files you want the next time you are on-line.

Logging On

When you first log onto a BBS, there are three pieces of information that you must provide: your first name, your last name, and your password (Fig. 2-2). If you have never called the particular BBS before, your name will not be recognized, so you will have an opportunity to choose a password of your own. When choosing password, remember the basic rules:

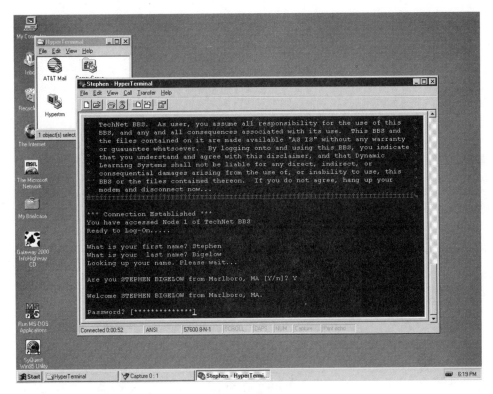

Figure 2-2 Logging on to a typical BBS.

- Choose a unique, nonobvious password. It should be something that you can remember easily, but prove difficult for someone else to guess. Do not use bank card numbers or other sensitive account numbers.
- Use at least five characters, with a mix of letters, numbers, and punctuation characters (if the BBS will allow it). The longer a password is, the more difficult it is to break or guess.
- Do not write the password down unless you can keep it in a secure place. Never reveal your password to others.
- If you use the BBS on a regular basis, make it a habit to change your password frequently.

The User Questionnaire

Once you have selected a password, most BBS facilities present you with a "user questionnaire" (or UQ). The UQ is typically a short, generic form that asks for mailing information, as well as some general information about your computer hardware. A UQ lets you establish your "account," and can be answered in less than five minutes. Generally speaking, completing a UQ is something of a courtesy; you can usually skip questions that you feel might be too personal. Once the UQ is finished, the system will check for e-mail, allow you to view the system newsletter, and take you to the main menu.

The System Newsletter

The wise *sysop* (system operator, pronounced sis'-op) understands that a BBS can provide an unwieldy amount of material for a user to sort through. The system newsletter (Fig. 2-3) allows the sysop to list new additions or current happenings on the system. For example, a newsletter might announce the addition of new files, new hardware, or new file areas. A newsletter also might announce deletions and file area closures. By keeping users abreast of the latest changes, they can take better advantage of the BBS. If you have the opportunity to view the newsletter, it is usually worthwhile reading.

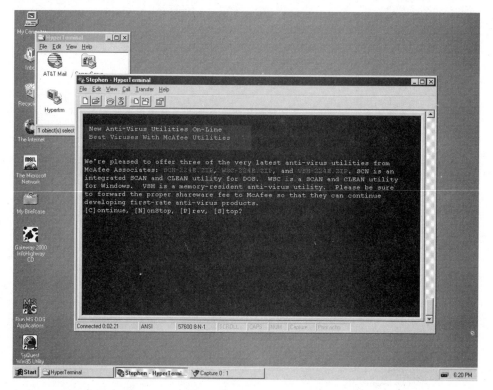

Figure 2-3 Reading the BBS on-line news.

Conferences Are King

A BBS primarily deals with files and e-mail, but grouping all files and e-mail together can be extremely cumbersome, rather like putting all the files on your computer in the root directory instead of using subdirectories. To help organize file areas and e-mail, the BBS uses *conferences* (Fig. 2-4). You can change your conference from the BBS main menu. By making various file areas available only from certain conferences, a BBS can be organized by topic area. For example, suppose there were file areas on amateur radio and file areas available on software engineering. The BBS might have an amateur radio conference (where only the ham-related file areas are available), and a software engineering conference (where only the engineering-related file areas are available). Some file areas might be available from more than one conference. When you post e-mail, your message is listed under the conference you are currently in. As an example, if you are in the hardware engineering conference, and you post an e-mail, the message will appear in that conference. So if someone else reads the messages in the software engineering conference, your post will not appear. By default, most BBS facilities have a "public" conference and a "general" conference.

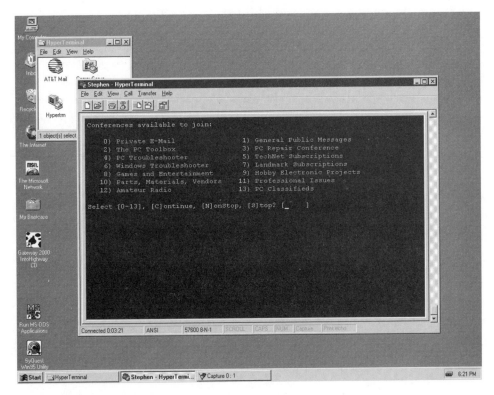

Figure 2-4 Checking for available conferences.

Notes on Uploading and
Downloading Files

Simply put, downloading is a process of copying a file from a BBS to your PC, while uploading is a process of copying a file from your PC to a BBS. Virtually all BBS facilities allow downloading, though many BBS systems will limit the number of files, the number of KB, or the amount of time you have available for downloading each day. For example, TechNet BBS limits new users to 10 files per day and 30 minutes per day of online time. Most BBS facilities solicit new contributions by allowing file uploads from users, but uploads are often limited to a few public or reserved areas. This gives the sysop a chance to check each contribution to see that it is appropriate and free of viruses before making them available in other file areas.

Time is a factor in file transfers; slower modems take longer to transfer a file. File transfer protocols also effect transfer times. Older protocols, such as Xmodem and Ymodem, take more time to transfer a file than newer protocols like Zmodem. If your communication software supports Zmodem, you should employ it whenever possible. Ultimately, faster modems and faster file transfer protocols speed your file transfers (and shorten the length of your toll call).

Finally, remember that many of the files available on a BBS are compressed using a file compression scheme like .ARJ or .ZIP. This is done to save space on the BBS hard drive(s) in order to hold more files, and because the compressed files are physically smaller, they can be downloaded faster. You need a utility to de-compress your downloaded files. Typically, the BBS has such utilities available for download also. Be sure to check any files for viruses before executing any demos or shareware.

E-mail

Posting an e-mail is a fairly straightforward matter; from the main menu, select the "message" menu (a new set of menu options will appear). You can then choose to read or post messages. If you choose to post a message, you simply type the message into a text window, then save the message. The BBS software takes care of all the header information for the message. Keep in mind that posting a message does not always guarantee a reply; depending on how busy the BBS is (and how busy the sysop is), a reply could take anywhere from several hours to several days.

NAVIGATING
COMMERCIAL SERVICES

A drawback to the BBS is its sense of isolation; each BBS is an "island" of information. Acquiring information from various different manufacturers required you to call several different BBS locations. Considering that every BBS has a different look and feel, the trend of "prospecting" for nuggets of information can quickly become tedious (and expensive). Around the time that private BBS setups were becoming popular, companies like CompuServe, GEnie, and America Online were starting their operations as *commercial* service providers.

Rather than single, independent computers, commercial providers linked multiple computers into a nation-wide (later international) network. Networking an information service provides several powerful advantages to commercial service providers. First, a network allows the system to be distributed across huge geographic areas, so users can typically find "local" access points without having to incur huge long-distance toll charges. Users can access the network for a relatively low hourly fee. Second, all of the resources of a network are cooperative, so a network can support many more users at the same time than a BBS can. Third, the nature of a network allows a level of redundancy, so the system can be more reliable than an individual BBS. It is interesting to note that most commercial on-line services now provide some level of access to the Internet.

Software

The software for a commercial on-line provider generally serves two purposes: it handles the communication between your modem and their network, and it provides the graphic "front end" such as the Windows dialogs and icons that are specific to the service provider (Fig. 2-5). Fortunately, the software used by commercial providers is usually given away for free; you can usually find the software for CompuServe and America Online packaged with issues of magazines like *PC Magazine* or *Windows Sources*. Instructions for installing and using the software are typically provided with the disk when you get it.

Logging On

Commercial services typically use a "dummy logon" approach for your initial subscription. In other words, your software disk comes with a unique "dummy" user name and password that allows you to log on in order to establish your account. Once your account is established, you can choose your own user name and password. For subsequent logons, the software might use a logon script, which automatically feeds your user name and password to the service (typical of CompuServe). Otherwise, you simply might have to enter your password when loging on (such as America Online).

Forums

Forums are to commercial services what conferences are to a BBS—they are a way of organizing information. For example, you can enter a "Game Publisher's

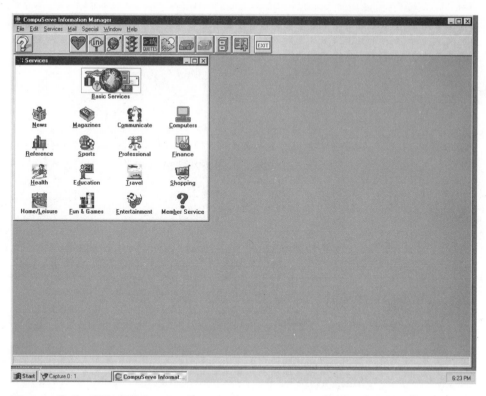

Figure 2-5   WinCIM—the CompuServe access software for Windows.

Forum," or the "PC Hardware Forum," or any one of hundreds of other forums that are available. When you enter a forum, there are up to ten or more file areas; each file area is managed by a particular manufacturer or organization (Fig. 2-6). It is this type of structure that has allowed hundreds of manufacturers to be so conveniently represented in an on-line service. Forums can be navigated through the use of a special icon in the on-line software. For example, CompuServe uses the "go" key (the traffic light icon), which prompts you for a keyword. America Online uses a "keyword" icon. Prodigy uses the "jump" command, and so on. You should remember that some forums command extra hourly fees.

Notes on E-mail

There are basically two types of e-mail in commercial online services: direct mail and forum messages. *Direct mail* is person-to-person mail; that is, you enter the ID of the individual you wish to send the message to, then compose your message in the corresponding dialog box (Fig. 2-7). If you wish to send me a message via CompuServe, you would address the message to *73652,3205*. If you want to send me a message from America Online, you would address the message to *SteveBige*. Forum messages are typically e-mail

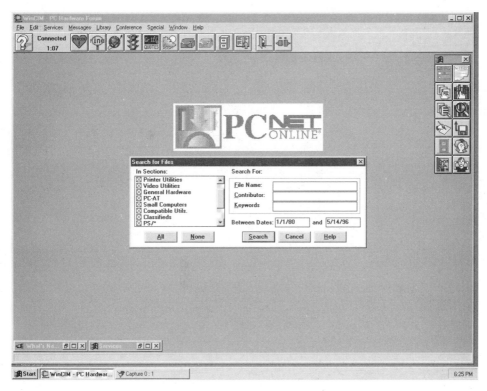

Figure 2-6   Checking the file areas in a selected forum.

postings to manufacturers within a forum (Fig. 2-8). For example, if you have a technical question about a product, and you are in that manufacturer's forum, you can post a question (or reply to someone else's posted question) within that forum. The big difference between direct mail and forum mail is that forum mail can be read by anyone in that forum, more of a communal message area. In most cases, e-mail is free. However, messages that are unusually long (or a high number of messages) might result in small surcharges.

USING THE INTERNET

Over the last few years, the Internet has finally come into its own as a resource for technicians and engineers alike. Manufacturers use the Internet for such diverse purposes as advertising, demo distribution, customer service, technical support, and even on-line order entry. A new generation of students are placing a wealth of data and circuit designs on-line. Professional organizations and extensive job-search sites are broadening the potential for job development. News sources are available to keep you abreast of the latest technical and business developments. Even private individuals are putting their experience and observations on the Internet. The broad appeal of the "information super-

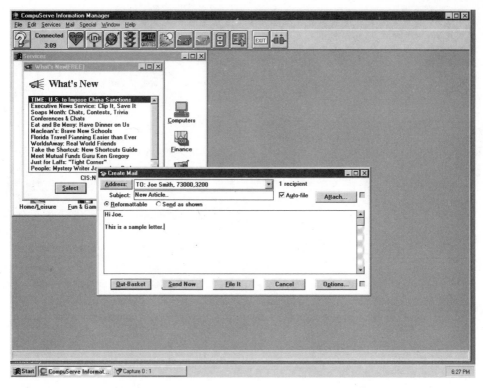

Figure 2-7 An example of "direct mail" through WinCIM.

highway" is attracting countless new users from around the world. From the convenience of a low-cost local service provider, you can access information from Boston to Sydney with equal ease.

Internet Software

Although there are a myriad of different functions available on-line, the Internet is best known for five main services: e-mail, the World Wide Web, newsgroups, mailing lists, and file transfer. The difference between Internet software and other types of on-line software is that Internet software is *modular* (each function is handled by a different utility) and there are several different utilities vying for dominance. For example, you would use a simple communication utility and TCP/IP stack to connect with your service provider (Fig. 2-9). Once the connection is established and running, you would then start the Internet program(s) to handle whatever functions you need (Fig. 2-10). If you wanted to explore the World Wide Web, you would start a "web browser" such as Netscape, Cello, or Mosaic (though Netscape is

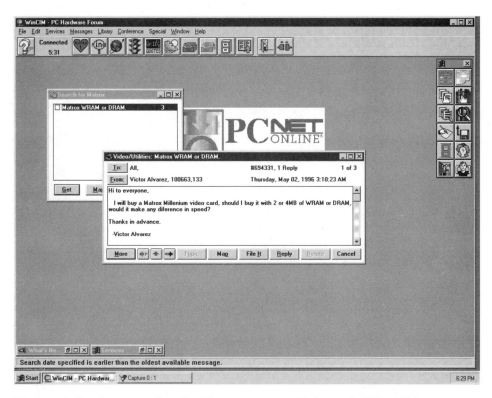

Figure 2-8 An example of a "forum message" through WinCIM.

certainly the most popular). If you wanted to send or receive e-mail, you would start a mail program like Eudora (pronounced U-dor'-a). When you decide to work with your newsgroups, you would start a program such as Trumpet (though Netscape also handles e-mail and newsgroups). As the Internet continues to develop, new and more powerful applications will appear.

This patchwork of Internet applications has its advantages and disadvantages. On the plus side, the competition between applications helps to ensure that users are not "stuck" with poor or bug-ridden software; you can choose the applications that best suit your needs. However, installing and using various pieces of software can be terribly confusing when compared to a single application (such as WinCIM, the CompuServe software).

Notes on Internet E-mail

Internet e-mail is remarkably similar to BBS or commercial service e-mail discussed earlier in this chapter. Once your e-mail package is running (i.e., Eudora), you simply select a "new message," then fill in the address informa-

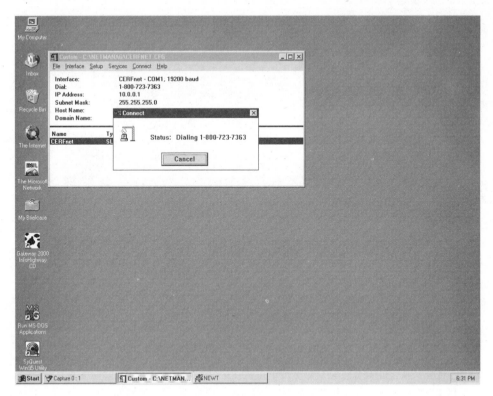

Figure 2-9   Connecting with an Internet provider.

tion and message text in the dialog box that appears (Fig. 2-11). For example, my Internet e-mail address is sbigelow@cerfnet.com; feel free to send mail once you finally get on-line.

One notable addition to Internet e-mail is the use of *signatures*. If you look at various pieces of e-mail, you might notice that many messages include the writer's name, address, telephone numbers, and other on-line contact information as shown in Fig. 2-12. In many cases, the contact information is wrapped in ornate ASCII graphics, which serve to personalize your message. The signature is a text file (which you can create) that is automatically added to your e-mail message; this eliminates the need for you to retype the signature each time.

Notes on the World Wide Web

One of the most popular features of the Internet has been the World Wide Web (often called "the web," "www," or "w3"). At its heart, the web is simply a series of hypertext files (known as HTML files), but the visual appeal of text and graphics, along with the ability to link countless pages via hypertext, has had stunning results. It is also possible to download files and send e-mail from hypertext links, as well as play sound and .AVI files. As a result, the web

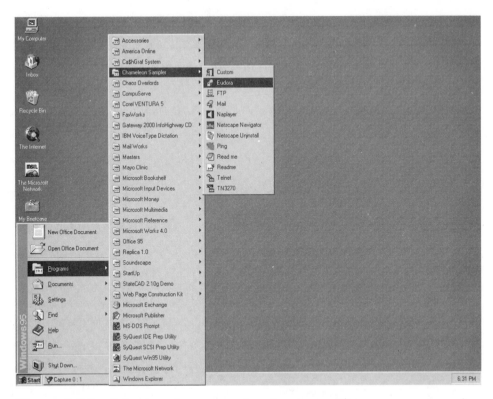

Figure 2-10   Selecting the appropriate Internet utility once a connection is made.

has quickly evolved into a flexible and exciting multimedia environment. This popularity is compounded by simplicity, because anyone can create interesting web pages in a relatively short time using little more than a word processor (though a wide range of powerful HTML authoring tools are now available).

Using the web is as simple as starting your web browser (Fig. 2-13) and entering the URL (*Universal Resource Locator*—the web page's address). Once the page is located, it is downloaded and displayed automatically. You can then read, save, or print the HTML file. As you read the document, notice that certain words or menu choices are highlighted. These are hypertext links that can take you to other places in the document, or other documents entirely, simply by clicking on the highlighted text. It is this ability to move quickly from one document to another that gave rise to the phrase "surf the web."

As you might expect, the problem with URLs is that they are just not intuitive. Having a web site has been likened to having a bulletin board in your living room, pretty useless unless someone knows it's there. Fortunately, there are now a myriad of web-based search engines available on the web that are designed to help you locate resources. Table 2-1 provides you with a list of the most popular search site URLs.

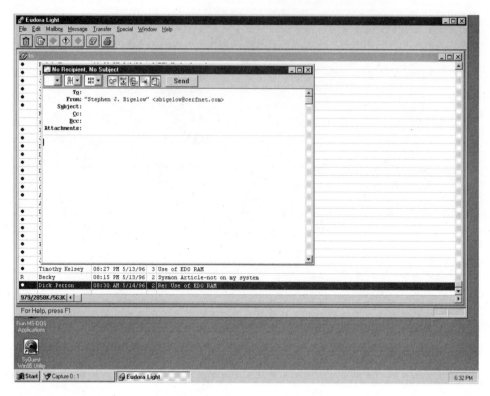

Figure 2-11   Composing Internet e-mail from Eudora.

There are also some points to keep in mind about URL commands and paths. For example, the command http://www.mysite.com/ is the root command; it will place you at the root directory of the site you are accessing. Often there is a default page (i.e., default.htm, home.htm, or index.htm) that will provide your links to other pages and subdirectories at the site. When a command and path are specified such as http://www.mysite.com/director/ actors/movies/movie.htm you still can try using the root command (http:// www.mysite.com/) to look for the "home page." This is a particularly useful trick when you don't find anything useful under the path you specified, but you want to examine a site further. Fortunately, web pages buried deep under layers of directories can be navigated quickly with any hypertext links that might be available on the page (i.e., "Click here to return to the home page").

This tactic also works with Gopher and FTP sites where paths to documents or software utilities are almost always specified. For example, if you don't find the document you need at ftp://ftp.mysite.com/PC/howto/repair.doc you can "back up" to ftp://ftp.mysite.com/PC/ and examine the other subdirectories you can choose from. Fortunately, Gopher and FTP browsers usually provide easy means for navigating the directories of remote sites.

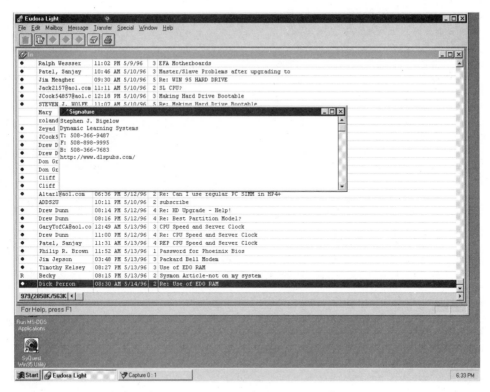

Figure 2-12   Examining the e-mail "signature."

Mailing Lists and News Groups

One of the main advantages of the Internet is its ability to bring together a vast number of people with similar interests; the problem is organizing all of those interests and providing an outlet for them all. The Internet handles this incredible diversity of interests and topics through the use of *mailing lists* and *news groups*. Each list or group has a particular (usually a very narrow) focus. By enrolling (or subscribing) to lists and groups of interest, you can participate in discussions and exchange messages with people of similar interests.

Mailing lists are handled through your e-mail utility. Messages from the list are automatically downloaded by your e-mail package, and you can post your own messages to the list through regular e-mail. As a result, no special software is needed to use mailing lists. Here are some points to remember about mailing lists:

- Subscribing to a mailing list is a simple matter; you need only send an e-mail to the mailing list server containing the command that will start your subscription (virtually all mailing lists are free). When sending a

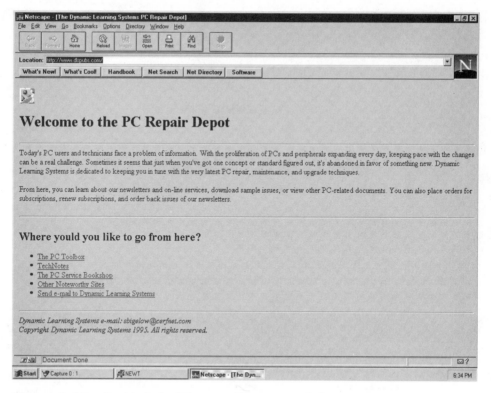

Figure 2-13 Starting the Netscape Internet web browser.

subscription command (or any command) to the list server, do not include a signature or any extraneous text in the body of the message. That will only confuse the list server. Once your subscription request is received, it might take from several minutes to several days to acknowledge the request and start your subscription. Busy lists might ask you to confirm your subscription request to verify your actual interest.

- When you first subscribe to a mailing list, you will receive a detailed e-mail outlining the rules of the list and any commands you can use (such as how to post messages to the list). Be sure to read the e-mail carefully and print a copy for your records. Failure to comply with the rules might get you bounced from the list.

- Stay on topic. Mailing lists are intended to be very specific in their intent and audience, and people don't have time to sort through e-mail that is not on their topic of interest. For example, a list on Windows 95 is not the right place to post your memory troubleshooting question. Posting an e-mail that is off-topic can result in a warning from the list moderator, flames from other list readers, and (if you persist) it might get you bounced from the list.

TABLE 2-1 Web Search Tools

Apollo	http://apollo.co.uk/
Architext	http://www.excite.com/
BizWeb	http://www.bizweb.com/
CityNet	http://www.city.net/forms/search.html
Commercial Searches	http://www.comcomsystems.com/search/
Commercial Sites	http://www.directory.net/
ElNnet Galaxy	http://galaxy.einet.net/
Global Commerce Net	http://www.commerce.com/
Global Village	http://www.homecom.com/global/global.htm
Harvest	http://harvest.cs.colorado.edu/
Information Super Lib.	http://www.mcp.com/
InfoSeek	http://www2.infoseek.com/
Inkotomi	http://204.161.74.6:1234/
Internet Mall	http://www.iw.com/imall/
Jump Station	http://js.stir.ac.uk/
Lycos	http://lycos-tmp1.psc.edu/
NCSA Mosaic What's New	http://gnn.com/gnn/wn/whats-new-html
Nerd World Media	http://www.nerdworld.com/users/dstein/index.html
Net Center	http://www.netcenter.com/netcentr/
Net Happenings	http://www.mid.net/directory.html
Net Search	http://www.infoseek.com/
Nikos	http://www.rns.com/nikos/
Open Text	http://www.opentext.com/
Pronet	http://www.pronett.com/
Starting Point	http://www.stpt.com/
Text Index	http://www.opentext.com:8080/
The Whole Internet	http://gnn.com/gnn/wic/index.html
The World	http://www.theworld.com/default.htm
Thomas Register	http://www.thomasregister.com/
Web Crawler	http://www.webcrawler.com/
WWW Virtual Library	http://www.w3.org/hypertext/DataSources/bySubject/Overview.html
WWW Worm	http://www.cs.colorado.edu/home/mcbryan/wwww.html
WWW Yellow Pages	http://www.yellow.com/
Yahoo	http://www.yahoo.com/

- Mail volume is another issue. "Hot" lists can generate dozens (sometimes hundreds) of messages each day. If you subscribe to one or more active lists, you can easily be deluged with hundreds of messages per day. Fortunately, you can usually suspend or withdraw from a list with a single command sent to the list server.

- Advertisements are strictly forbidden; in the Internet community, that is called *spamming*, and will get you removed from a list almost immediately (and usually permanently). If you have something to sell, look for lists specifically set up for buy-and-sell operations.
- Lists are either moderated or unmoderated. In a moderated list, someone reviews each message before it is posted to the list. This often results in a higher quality list that is more on-topic, but might take a bit longer for messages to hit the list. An unmoderated list has little (if any) human intervention, and can easily become off-topic or erupt with flame wars, which often reduces the list's quality.
- Most mailing lists are *continuous;* messages are distributed to the list as soon as they are received and cleared. However, you will encounter some *periodical* lists that are distributed on a weekly, biweekly, or monthly basis.
- Don't be surprised to see mailing lists (especially moderated lists) change and frequently disband because only one or two people bear the burden of list management.

News groups are remarkably similar to mailing lists in their narrow focus and general rules of behavior, but the process of using a news group is a bit different. The most notable difference is that news groups operate through a *news reader* utility. Trumpet is a popular shareware news group utility, but Netscape also will serve as a news group reader (Fig. 2-14). You enter the news group identifier (i.e., comp.pc.ibm.repair) into the reader and subscribe. At that point, you can access the messages posted in the group, read them at will, answer any questions or issues of interest, and post "articles" of your own. You also can quit your subscription through the news group reader. As a result, news group messages do not "download" themselves into your PC as list messages do. With news groups, you are interacting with the news group server directly.

As with web sites, mailing lists and news groups are often hidden in a nether-world of arcane terms and acronyms. Several search engines have evolved to help you locate meaningful lists and groups. Table 2-2 lists a number of the more popular search engines.

Notes on FTP

Although your web browser will support file downloads from web pages, there are many circumstances where files are made available through *FTP* (File Transfer Protocol). Basically, FTP transfers are made by starting the FTP utility (Fig. 2-15), entering the FTP location and your access information, then navigating the directories of the remote computer until you find the file(s) of

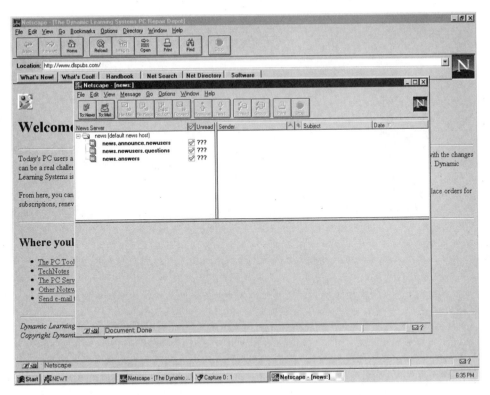

Figure 2-14   Using a news reader from within Netscape.

interest. In effect, you can explore the directories of the remote computer as if it were just another drive on your PC. You then can transfer files from the remote computer to your PC. If you have the proper access, you also can upload files to the remote computer.

FTP access is an important security issue. Most FTP sites are protected by a user name and password that you must have before access to a site will be permitted; this is the "holy grail" for computer hackers. For general-access FTP sites, however, you usually can accomplish an *anonymous* logon using your e-mail name as the User Name, and the word "anonymous" as the password. This will generally give you access to any public areas that might be available.

TABLE 2-2   Mailing List/News Group
Search Engines

http://www.intserv.com/~lbi
http://catseye.bluemarble.net/
http://www.dejanews.com/

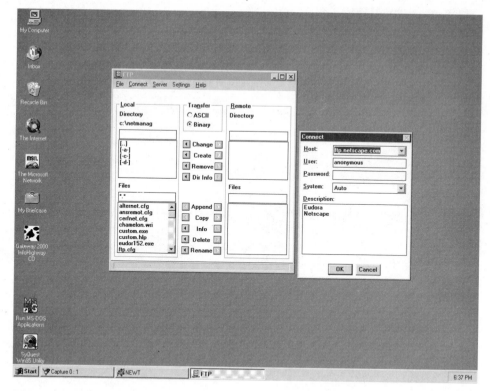

Figure 2-15 Using an FTP utility to transfer files.

The speed of your file transfer is dependent on your modem speed, as well as the processing demands on the particular FTP site. For example, even a fast modem or ISDN connection can be bogged down by an overloaded FTP server. As a result, you should put off large file transfers until the evenings or weekends when FTP site is less congested.

Viruses and the Internet

The Internet has emerged as a major resource for software distribution. Unfortunately, the lack of central management (and lackluster virus checking on the part of some software distributors) also has made the Internet a prime source for the distribution of computer viruses. Now in practice, there is only a very small instance of virus damage from downloaded files, but the potential is certainly there. As a rule, you should check all software downloaded from any source before running "setup" or executing the software for the first time. Also make sure that you are running the latest available version of a virus checker.

CHAPTER

3

On-line Resources
A through Z

CORPORATE/
MANUFACTURER
CONTACTS

20/20 Software (*SW Utilities,*
Development)
8196 SW Hall Blvd., Suite 200
Beaverton, OR 97008
T: 503-520-0504
F: 503-520-9118
CIS: go TWENTY (or 74774,222)
E-mail: info@twenty.com
URL:
http://www.twenty.com/~twenty/

3Com Corporation (*Networking,*
Hardware)
5400 Bayfront Plaza
P.O. Box 58145
Santa Clara, CA 95052-8145
T: 408-764-5000
F: 408-764-6740
URL: http://www.3com.com/

3DO (*Computers, Dedicated*)
The 3DO Company
600 Galveston Dr.
Redwood City, CA 94063
T: 415-261-3000
Service: 415-261-3454
URL: http://www.3do.com/

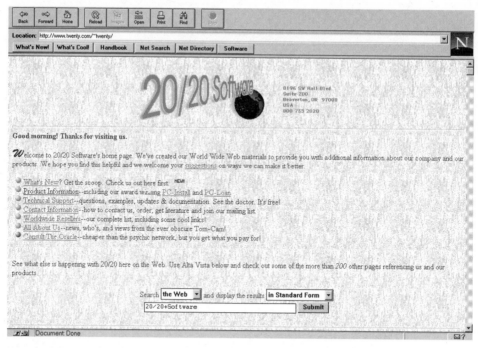

Figure 3-1 20/20 Software.

3M Data Storage Products (*Peripherals, Data Storage*)
3M Center Building, #223-5N-01
St. Paul, MN 55144
T: 612-736-1866
F: 612-736-1246
URL: http://www.mmm.com/
product/opunits.html#datastorage

A2Z Computers (*Distributor, PC Assemblies*)
701 Beta Dr., Unit 19
Mayfield Village, OH
T: 800-983-8889
URL: http://www.a2zcomp.com/

Aaeon (*Equipment, Test & Programming*)
274 Raritan Center Parkway
Raritan Center
Edison, NJ 08837
T: 908-346-0051
F: 908-346-0011
E-Mail: aaeon@webleader.com.
URL: http://www.aaeon.com/

Aavid Thermal Technologies, Inc.
(*Thermal Products, Heat Sinks*)
One Kool Path
P.O. Box 400
Laconia, NH 13247-0400
T: 603-528-3400
F: 603-527-2129
URL: http://www.aavid.com/

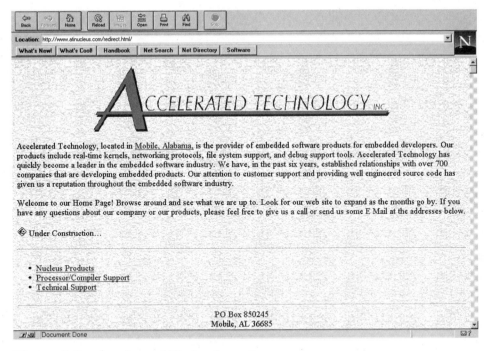

Figure 3-2 Accelerated Technology, Inc.

Abekas Video Systems (Scitex) (*Video, Multimedia Equipment*)
Scitex Digital Video
101 Galveston Dr.
Redwood City, CA 94063
T: 415-369-5111
F: 415-369-4777
E-mail: info@scitexdv.com
URL: http://www.abekas.com/

Abstract Technologies (*Peripherals, RISC Workstations*)
4032 South Lamar Blvd., Suite 500-142
Austin, TX 78704-7900
T: 512-441-4040
F: 512-416-0310
E-mail: info@abstract.com
URL: http://www.abstract.co.nz/

ACC (Advanced Computer Communication) (*Networking, Hardware*)
340 Storke Rd.
Santa Barbara, CA 93117
T: 805-685-4455
F: 805-685-4465
E-mail: info@acc.com
URL: http://www.acc.com/

Accelerated Technology, Inc. (*SW Utilities, Embedded Product Development*)
PO Box 850245
Mobile, AL 36685
T: 334-661-5770
F: 334-661-5788
E-mail: atisales@atinucleus.com
URL: http://www.atinucleus.com/

ACC (Associated Computer Consultants)   (*Networking, Hardware*)
8320 Guilford Rd., Suite G
Columbia, MD 21046
T: 800-242-0739
F: 410-290-8106
URL: http://www.sys.acc.com/

Accton   (*Networking, Hardware*)
Accton Technology Corporation (USA)
1962 Zanker Rd.
San Jose, CA 95112
T: 408-452-8900
F: 408-452-8988
B: 408-452-8828
FaxBack: 408-452-8811
URL: http://www.accton.com/

Acer America Corporation
(*Computers, Systems*)
399 West Trimble Rd.
San Jose, CA 95131
T: 800-733-2237
Tech Sup: 800-445-6495
B: 408-428-0140 (28.8Kbps, 8-N-1)
AOL: keyword ACER
Prodigy: jump ACER
CIS: go ACER
FTP: ftp://ftp.acer.com
URL: http://www.acer.com/

ACIS, Inc.   (*Distributor, Memory*)
2381 Philmont Ave.
Huntingdon Valley, PA 19006
T: 215-938-4291
F: 215-938-4290
URL: http://www.aciscorp.com/

Actel Corporation   (*ICs, Logic*)
955 East Arques Ave.
Sunnyvale, CA 94086-4533
T: 800-228-3532
F: 408-739-1540
URL: http://www.actel.com/index.html

ADAC   (*Peripherals, Data Acquisition*)
70 Tower Office Pk.
Woburn, MA 01801
T: 617-935-3200
F: 617-932-3200
E-mail: info@adac.com
URL:
http://www.adac.com/adac/index.html

Adaptec   (*Controllers, Drive & Network*)
Technical Support, M/S 105
691 South Milpitas Blvd.
Milpitas, CA 95035
T: 408-945-8600
Technical Sup: 800-959-7274
Technical Sup: 408-934-7274
FaxBack: 408-957-7150
Literature Hotline: 510-732-3829
Software Hotline: 408-957-7274
B: 408-945-7727 [8 data bit, 1 stop bit,
no parity, 28,800 bps supported]
E-mail: support@adaptec.com
FTP: ftp.adaptec.com
URL: http://www.adaptec.com/

Adcom   (*Peripherals, Sound*)
11 Elkins Rd.
East Brunswick, NJ 08816
T: 908-390-1130
E-mail: adcom@soundsite.com
URL:
http://www.soundsite.com/adcom/

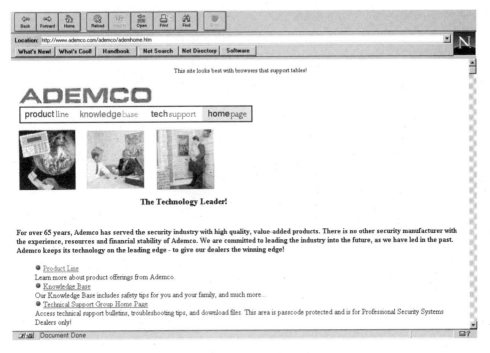

Figure 3-3 Ademco.

ADC Telecommunications
(*Networking, Enterprise*)
12501 Whitewater Dr.
Minnetonka, MN 55343
T: 612-938-8080
Technical Sup: 800-366-3891
E-mail: technical@adc.com
URL: http://www.adc.com/

Addison-Wesley Publishing   (*Publisher,*
Computer)
One Jacob Way
Reading, MA 01867
T: 617-944-3700
F: 617-944-9338
URL: http://www.aw.com/

Ademco   (*Security, Hardware*)
165 Eileen Way
Syosset, NY 11791
T: 800-645-7568
URL: http://www.ademco.com/

Adept Scientific   (*Peripherals, Data*
Acquisition)
6 Business Centre West, Avenue One
Letchworth, Hertfordshire, SG6 2HB UK
T: +44 1462 480055
F: +44 1462 480213
E-mail: info@adeptscience.co.uk
E-mail: support@adeptscience.co.uk
URL: http://www.adeptscience.co.uk/

Adobe Systems (*SW Utilities, DTP & Imaging*)
1585 Charleston Rd.
P.O. Box 7900
Mountain View, CA 94039-7900
T: 415-961-4400
F: 415-961-3769
B: 206-623-6984
FaxBack: 206-628-5737
FTP: ftp://ftp.adobe.com/
URL: http://www.adobe.com/

Adtran (*Communications, Hardware & ISDN*)
901 Explorer Blvd.
Huntsville, AL 35806-2807
T: 800-827-0807
B: 205-971-8169
E-mail: support@adtran.com
E-mail: cservice@adtran.com
URL: http://www.adtran.com/

Advanced Digital Information Corp.
(*Peripherals, Tape*)
14737 NE 87th St.
Box 2996
Redmond, WA 98073-2966
T: 206-881-8004
F: 206-881-2296
B: 714-894-0893

Advanced Digital Systems (*Video, PC-TV Conversion*)
13909 Bettencourt St.
Cerritos, CA 90703
T: 310-926-1928
F: 310-926-0518
URL: http://www.ads-mm.com/

Advanced Hardware Architectures
(*Tools, Data Compression & Error Correction*)
2365 NE Hopkins Ct.
Pullman, WA 99163-5601
T: 509-334-1000
F: 509-334-9000
E-mail: sales@aha.com
URL: http://www.aha.com/

Advanced Logic Research (ALR)
(*Computers, Systems*)
9401 Jeronimo
Irvine, CA 92718
T: 714-581-6770
F: 714-458-0532
Technical Support: 800-257-1230
BBS: 714-458-6834
CIS: go ALRINC
E-mail: tech@alr.com
URL: http://www.alr.com/

Advanced Microcomputer Systems
(*Simulation, Circuit CAD*)
1460 S.W. 3rd St.
Pompano Beach, FL 33069
T: 305-784-0900
F: 305-784-0904
E-mail: ams@gate.net
URL: http://www.gate.net/~ams/

Advanced Micro Devices (AMD)
(*ICs, Microprocessors*)
One AMD Place
P.O. Box 3453
Sunnyvale, CA 94088
T: 408-732-2400
URL: http://www.amd.com/

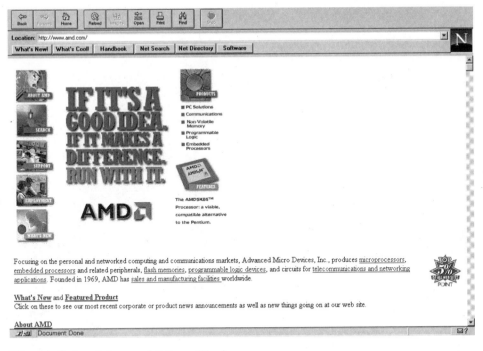

Figure 3-4   Advanced Micro Devices.

Advanced Microelectronics
(*ICs, Custom Design*)
1080 River Oaks Dr., Suite A-250
Jackson, MS 39208
T: 601-932-7620
F: 601-932-7621
E-mail: design@aue.com
URL:
http://www.aue.com/aue_home.html

Advanced Quick Circuits   (*Manufacturer, PCB*)
245 East Dr.
Melbourne, FL 32904
T: 407-768-9901
F: 407-725-6595
URL: http://www.iu.net/aqc/

Advanced RISC Machines   (*ICs, RISC Processors*)
985 University Ave., Suite 5
Los Gatos, CA 95030
T: 408-399-5199
F: 408-399-8854
E-mail: info@armltd.co.uk
FTP: ftp://ftp.acorn.co.uk/
URL: http://www.arm.com/

Advanced Storage Concepts
(*Controllers, SCSI*)
10713 Ranch Rd. 620 N., Suite 305
Austin, TX 78726
T: 512-335-1077
F: 512-335-1078
B: 512-335-3499
E-mail: asc@eden.com
URL: http://www.eden.com/~asc/

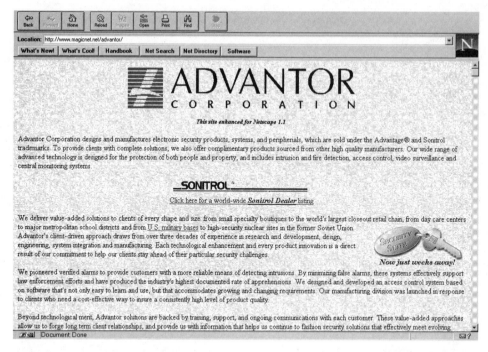

Figure 3-5 Advantor Corporation.

Advanced Systems (*Computers, Dedicated*)
19 Portland Square
Bristol BS2 8SJ UK
T: +44 (0)117 9427219
F: +44 (0)117 9241425
E-mail: support@asl.com
URL: http://www.asl.com/

Advance International (*Equipment, Power Supplies*)
Kidd House
Sandbeck Way
Wetherby, West Yorkshire, LS22 7DD UK
T: +44 1937 581961
F: +44 1937 586908
URL: http://www.dircon.co.uk/advance-int/

AdvanSys (*Controllers, Drive*)
1150 Ringwood Ct.
San Jose, CA 95131
T: 408-383-9400
F: 408-383-9612
B: 408-383-9540
FaxBack: 408-383-9753
URL: http://www.advansys.com/

Advantor Corporation (Sonitrol)
(*Security, Hardware*)
E-mail: barney@advantor.com
URL: http://www.magicnet.net/advantor/ (Includes contacts and dealer listings)

Advent Electronics (*Equipment, Cable & Satellite*)
10615 Heatherford Ct.
Houston, TX 77041
T: 713-466-9994
URL:
http://www.intergate.com/~mhammer/

AEG Schneider Automation (US)
(*Equipment, PLC*)
One High St.
North Andover, MA 01845
T: 508-794-0800
F: 508-975-9010
URL: http://www.modicon.com/

Agile Networks (*Networking, Hardware*)
1300 Massachusetts Ave.
Boxborough, MA 01719
T: 800-ATM-9LAN
F: 508-263-5111
E-mail: info@agile.com
URL: http://www.agile.com/

Ainsworth Technology (*Networking, Hardware*)
131 Bermondsey Road
Toronto, Ontario, M4A 1X4 Canada
T: 416-751-4420
F: 416-751-9031
URL:
http://wchat.on.ca/ainsworth/ati.htm

Aladdin Knowledge Systems, Ltd.
(*SW Utilities, Development*)
The Empire State Building
350 Fifth Ave., Suite 6614
New York, NY 10118
T: 212-564-5678
F: 212-564-3377
E-mail: sales@us.aks.com
URL: http://www.hasp.com/

Alantec (*Networking, Hardware*)
2115 O'Nel Dr.
San Jose, CA 95131
T: 408-955-9000
F: 408-955-9500
URL: http://www.alantec.com/product/product.html

Alaris (*Computers, Motherboards*)
411 1st Ave. S.
Seattle, WA 98104
T: 206-622-5500
F: 206-343-3360
B: 206-623-6984

Alberta Printed Circuits, Ltd.
(*Manufacturer, PCB*)
#3, 1112 - 40th Ave. N.E.
Calgary, Alberta, T2E 5T8 Canada
T: 403-250-3406
B: 403-291-9342
E-mail: staff@apcircuits.com
FTP: ftp://ftp.apcircuits.com/
URL: http://www.apcircuits.com/

Aldec, Inc. (*Simulation, Circuit CAD*)
3 Sunset Way, Suite F
Henderson, NV 89014
T: 702-456-1222
F: 702-456-1310
URL: http://www.aldec.com/

Alden Electronics, Inc. (*Equipment, Weather & Imaging*)
40 Washington St.
P.O. Box 500
Westboro, MA 01581-0500
T: 508-366-8851
F: 508-898-2427
E-mail: info@alden.com
E-mail: help@alden.com
URL: http://www.alden.com/

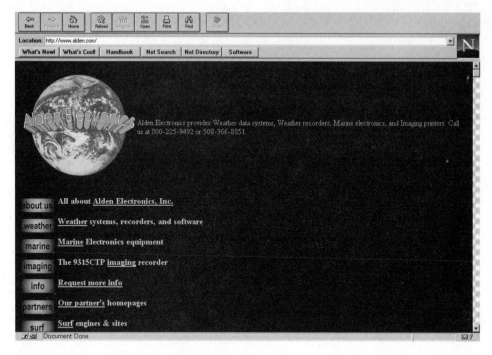

Figure 3-6   Alden Electronics, Inc.

Aliroo Ltd.   (*SW Utilities, Encryption*)
19 Trumpeldur St. Kefar-Sava 44442
Israel
T: 972-9-7677732
F: 972-9-7677739
E-mail: aliroo@netvision.net.il
URL:
http://www.netvision.net.il/~aliroo/

ALL Computer   (*Peripherals, Memory*)
1220 Yonge St., 2nd Floor
Toronto, Ontario, M4T 1W1 Canada
T: 416-960-0111
F: 416-960-5426
B: 416-960-8679

Allegro New Media   (*SW Utilities,
Business & Reference*)
16 Passaic Ave., Suite 6
Fairfield, NJ 07004
T: 201-808-1992
F: 201-808-2645
CIS: 70007,5145
AOL: keyword AllegroNMe
Prodigy: jump NYGY14E
URL: http://www.allegronm.com/

Allen-Bradley (Rockewell Automation)
(*Equipment, Industrial*)
URL: http://www.ab.com/   (Includes
worldwide contacts index)

Allied Electronics (*Distributor, Components*)
7410 Pebble Dr.
Ft. Worth, TX 76118-6997
T: 817-595-3500
F: 817-595-6444
URL: http://www.allied.avnet.com/
(Includes worldwide contacts index)

AllMicro (*Diagnostics, Hardware & Software*)
18820 US Hwy. 19 N., #215
Clearwater, FL 34624
T: 813-539-7283
F: 813-531-0200
B: 813-535-9042
E-mail: allmicro@allmicro.com
URL: http://www.allmicro.com/

Alloy Computer Products
(*Peripherals, Tape*)
25 Polter Rd.
Littleton, MA 01460
T: 508-486-0001
F: 508-486-3755
B: 508-486-4044

Alpha Telecom, Inc. (Vector Technology) (*Communications, Hardware & ISDN*)
15111 Mintz Ln.
Houston, TX 77014-1400
T: 713-440-8340
F: 713-440-8460
E-mail: techsupport@vector.com
URL:
http://www.vector.com/alpha.html

Alps Electric (USA) (*Computers, Systems*)
T: 800-449-2577
B: 408-432-6424

Alta Technology (*Peripherals, Supercomputers*)
9500 South 500 West, Suite 212
Sandy, UT 84070-6655
T: 801-562-1010
F: 801-254-2020
E-mail: support@altatech.com
URL: http://www.xmission.com/~altatech/

Altec Lansing Corporation (*Peripherals, Speakers*)
P.O. Box 26105
Oklahoma City, OK 73126
T: 405-324-5311
F: 405-324-8981
E-mail: 75533.2171@compuserve.com
URL:
http://www.sysint.com/systems/altec/

Altera Corporation (*ICs, Logic*)
2610 Orchard Pkwy.
San Jose, CA 95134-2020
T: 408-894-7900
F: 408-433-3943
CIS: go ALTERA
E-mail: sos@altera.com
FTP: ftp://ftp.altera.com
URL: http://www.altera.com/

Alternate Source Components
(*Distributor, Components*)
1575 Military Rd., Unit 13-170
Niagara Falls, NY 14304-4706
T: 905-825-3044
F: 905-825-4132
URL: http://www.alternatesrc.com/

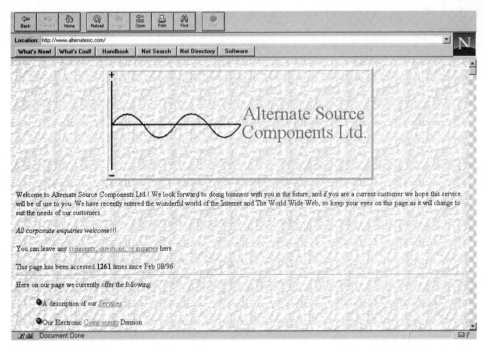

Figure 3-7 Alternate Source Components.

Amber Wave Systems (*Networking, Hardware*)
42 Nagog Park
Acton, MA 01720-3409
T: 508-266-2900
F: 508-266-1159
E-mail: info@amberwave.com
FTP: ftp://ftp.amberwave.com/
URL: http://www.amberwave.com/

AMCC (*ICs, Logic*)
25 Burlington Mall Rd., Ste. 300
Burlington, MA 01803
T: 617-270-0643
F: 617-221-5853
E-mail: danc@amcc.com
URL: http://www.amcc.com/

Amdahl (*Computers, Servers*)
1250 East Arques Ave.
Sunnyvale, CA 94088-3470
T: 408-746-6000
F: 408-746-3243
URL: http://www.amdahl.com/

Amdek Corporation (*Peripherals, Monitors*)
3471 N. First St.
San Jose, CA 95134
T: 408-473-1200
F: 408-922-5729
B: 408-922-4400

Amecon, Inc. (*Computers, Hardware & Multimedia*)
10 Ferry Wharf
Newburyport, MA 01950

T: 508-463-7469
F: 508-463-7480
E-mail: sales@amecon.com
URL: http://www.amecon.com/

America Online   (*Services, On-line*)
8619 Westwood Center Dr.
Vienna, VA 22182
T: 703-448-8700
URL: http://www.blue.aol.com/

American Cybernetics   (*SW Utilities, Internet*)
1830 W. University Dr., Ste 112
Tempe, AZ 85281
T: 602-968-1945
F: 602-966-1654
CIS: go CYBERNET
E-mail: sales@amcyber.com
E-mail: tech@amcyber.com
URL: http://www.amcyber.com/

American Megatrends, Inc. (AMI)
(*BIOS Motherboards, Diagnostics*)
6145-F Northbelt Parkway
Norcross, GA 30071
T: 770-246-8600
F: 770-246-8790
B: 404-246-8780
FaxBack: 770-246-8787
FTP: ftp://ftp.megatrends.com/ftp/
URL: http://www.megatrends.com/

American Micro Products Technology
(*Distributor, Motherboards*)
AMPTech Corporate Center
5351 Naiman Parkway
Solon, OH 44139
T: 216-498-9564
E-mail: amptech@icgroup.net
URL: http://www.amptech.com/

American Micro Solutions (AMS)
(*Distributor, PC Assemblies*)
15461 Redhill Ave., Ste E
Tustin, CA 92680
T: 714-258-8818
F: 714-258-8918
E-mail: ams@mail.calypso.com
URL: http://www.calypso.com/ams/

American Power Conversion (APC)
(*Equipment, Power Products*)
132 Fairgrounds Rd.
W. Kingston, RI 02892
T: 401-789-5735
F: 401-789-3180
FaxBack: 800-347-3299
CIS: go APCSUPPORT
URL: http://www.apcc.com/

AMP, Inc.   (*Computers, Connectors*)
Harrisburg, PA 17105-3608
T: 717-986-7777
F: 717-986-7575
FaxBack: 717-986-3500
URL: http://www.amp.com/

Analog Devices   (*ICs, Analog*)
1 Technology Way
PO Box 9106
Norwood, MA 02061
T: 617-329-4700
F: 617-326-8703
FTP: ftp://ftp.analog.com/
URL: http://www.analog.com/

AnaTek Corporation   (*Diagnostics, Software*)
PO Box 1200
Amherst, NH 03031
T: 800-999-0304
F: 603-673-5374
E-mail: info@anatekcorp.com
URL: http://anatekcorp.com/repair.htm

Annabooks (*Publisher, Computer*)
11838 Bernardo Plaza Court
San Diego, CA 92128-2414
T: 619-673-0870
F: 619-673-1432
E-mail: info@annabooks.com
URL: http://www.annabooks.com/

Ansoft Corporation (*Simulation, Circuit CAD*)
Four Station Square, Ste. 660
Pittsburgh, PA 15219-1119
T: 412-261-3200
F: 412-471-9427
EMAIL: info@ansoft.com
URL: http://www.ansoft.com/

Antares Corporation (*Simulation, Circuit CAD*)
8905 S.W. Nimbus Ave., Suite 155
Beaverton, OR 97008
T: 503-643-5800
E-mail: info@antaresco.com
URL: http://www.antaresco.com/

Antec, Inc. (*Computers, Hardware*)
2859 Bayview Dr.
Fremont, CA 94538
T: 510-770-1200
F: 510-770-1288
E-mail: antec@antec-inc.com
URL: http://www.antec-inc.com/

Apem (*Manufacturer, Components*)
134 Water St.
P.O. Box 544
Wakefield, MA 01880-4444
T: 617-246-1007
F: 617-245-4531
E-mail: info@apem.com
URL: http://www.apem.com/

Apple Computer (*Computers, Systems*)
1 Infinite Loop
Cupertino, CA 95014
T: 408-996-1010
FaxBack: 800-505-0171
URL: http://www.apple.com/

Applied Innovation, Inc. (*Equipment, Cable*)
5800 Innovation Dr.
Dublin, OH 43017
T: 614-798-2000
URL:
http://invest.quest.columbus.oh.us:80/
InvestQuest/a/ainn/

Applied Microsystems Corporation
(*SW Utilities, Development*)
5020 148th Ave. NE
Redmond, WA 98073-9702
T: 800-275-4262
F: 206-883-3049
E-mail: support@amc.com
URL: http://www.amc.com/

APPRO International, Inc.
(*Computers, Industrial*)
2032 Bering Dr.
San Jose, CA 95131
T: 408-452-9200
F: 408-452-9210
E-mail: appro@appro.com
URL: http://www.appro.com/

APS Technologies (*Distributor, Drives*)
6131 Deramus
Kansas City, MO 64120
T: 800-947-8125
URL: http://www.apstech.com/

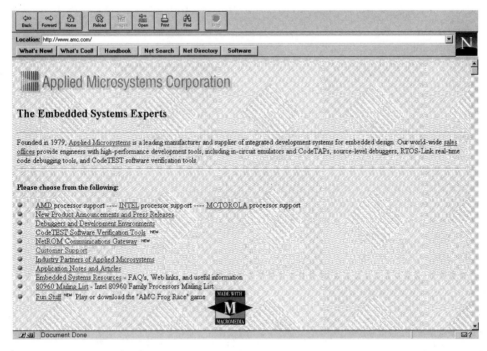

Figure 3-8   Applied Microsystems Corporation.

Arco Electronics, Inc.   (*Controllers, Drive*)
2750 N. 29th Ave., Suite 316
Hollywood, CA 33020
T: 305-925-2688
F: 305-925-2889
B: 305-925-2791

Argus Technologies, Ltd.   (*Equipment, Power Products*)
7033 Antrim Ave.
Burnaby, British Columbia, V5J 4M5
Canada
T: 604-436-5900
F: 604-436-1233
URL: http://www.nstn.ca/argus/

Ariel Corporation   (*Peripherals, DSP*)
2540 Route 130
Cranbury, NJ 08512-3507
T: 609-860-2900
F: 609-860-1155
URL: http://www.ariel.com/default.htp

Aries Electronics   (*Computers, Connectors*)
URL: http://www.arieselec.com/
(Includes links to worldwide distributors)

Aris Microsystems (*Computers, Systems*)
24 New Industrial Rd., #06-02 Pei Fu
Industrial Building
Singapore 1953
T: 65-285-7161
Tech Support: 65-220-1070
F: 65-285-6876
URL:
http://www.aris.com.sg/aris/aris.html

Aromat Corporation (Matsushita)
(*Manufacturer, Components*)
629 Central Ave.
New Providence, NJ 07974
T: 908-464-3550
F: 908-464-8513
URL: http://www.mew.com/index.html_
Aromat

Arrick Robotics (*Automation,
Robotics*)
PO Box 1574
Hurst, TX 76053
T: 817-571-4528
F: 817-571-2317
E-mail: info@robotics.com
URL:
http://www.robotics.com/index.html

Asante Technologies, Inc. (*Networking,
Hardware & Software*)
T: 408-435-8388
F: 408-432-6018
B: 408-432-1416
FaxBack: 408-954-8607
E-mail: support@asante.com
FTP: ftp://ftp.asante.com/
URL: http://www.asante.com/

Ascend Communications (*Networking,
Hardware*)
1275 Harbor Bay Parkway
Alameda, CA 94502

T: 510-769-6001
F: 510-814-2300
E-mail: info@ascend.com
E-mail: support@ascend.com
URL: http://www.ascend.com/

askSam Systems (*SW Utilities,
Databases*)
PO Box 1428
Perry, FL 32347-9968
T: 904-584-6590
B: 904-584-5413
CIS: go ASKSAM
E-mail: tech@asksam.com
URL: http://www.askSam.com/

Aslan Computer (*Equipment, Power
Supplies*)
5101 Commerce Dr.
Baldwin Park, CA 91706
T: 818-337-6860
F: 818-960-1690
URL: http://www.gus.com/emp/aslan/
aslan.html

Aspen Systems, Inc. (*Computers,
Systems*)
4026 Youngfield St.
Wheat Ridge, CO 80033-3862
T: 303-431-4606
F: 303-431-7196
E-mail: aspen@aspsys.com
URL: http://www.aspsys.com/

Astec (*Equipment, Power Products*)
6339 Paseo del Lago
Carlsbad, CA 92009
T: 619-757-1880
F: 619-930-4700
FaxBack: 619-930-0881
URL: http://www.astec.com/

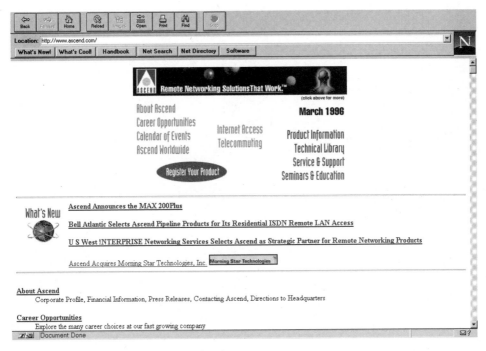

Figure 3-9 Ascend Communications.

AST Research (*Computers, Systems*)
16215 Alton Parkway
Irvine, CA 92718
T: 714-727-4141
F: 714-727-9355
B: 714-727-4723
URL: http://www.ast.com/

Asus (*ICs, BIOS*)
1635 McCandless Dr.
Milpitas, CA 95035
T: 408-956-9077
F: 408-956-9088
B: 408-956-9084
E-mail: info-usa@asustek.asus.com.tw
E-mail: tsd-usa@asustek.asus.com.tw
URL: http://asustek.asus.com.tw/

Atec, Inc. (*Manufacturer, Custom*)
12600 Executive Dr.
Stafford, TX 77477
T: 713-240-1919
E-mail: flex@atec.com
URL:
http://www.sccsi.com/Atec/home.html

ATI Technologies (*Controllers, Video*)
33 Commerce Valley Dr. E
Thornhill, ON, L3T 7N6, Canada
T: 905-882-2600
F: 905-882-2620
B: 905-764-9404
CIS: go ATITECH
CIS: 74740,667
FTP: ftp.gateway.centre.com
URL: http://www.atitech.ca/

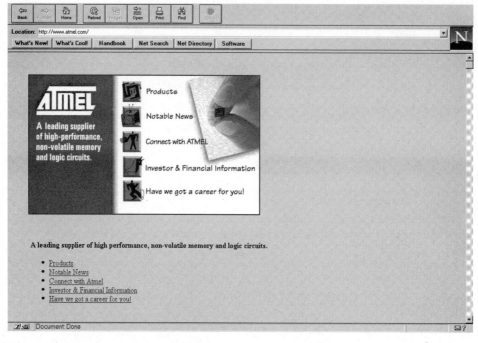

Figure 3-10 Atmel Corporation.

Atlas Soundolier (*Peripherals, Sound*)
1859 Intertech Dr.
Fenton, MS 63026
T: 314-349-3110
F: 314-349-1251
URL:
http://www.sysint.com/systems/atlas/

Atmel Corporation (*ICs, Logic*)
2125 O'Nel Dr.
San Jose, CA 95131
T: 408-436-4270
F: 408-436-4314
URL: http://www.atmel.com/

Attachmate (*SW Utilities, Internet*)
3617 131st Ave. SE
Bellevue, WA 98006

T: 206-644-4010
F: 206-747-9924
URL: http://www.atm.com/

AT&T (*Communications, Hardware*)
2551 E. 40th Ave.
Denver, CO 80205 (National Parts)
T: 800-222-7278
Tech Support: 800-628-2888
Support Fax: 800-527-4360
URL: http://ncrinfo.attgis.com/
(NCR information)
URL: http://www.att.com/
(Includes all directories and indexes for reference)
URL: http://www.ncr.com/

Auspex Systems, Inc. (*Computers, Servers*)
5200 Great America Parkway
Santa Clara, CA 95054
T: 408-986-2000
F: 408-986-2020
E-mail: info@auspex.com
URL: http://www.auspex.com/

Austin Computers (*Computers, Systems*)
2121 Energy Drive
Austin, TX 78758
T: 512-339-3500
F: 512-454-1357
URL: http://www.ipctechinc.com/

Autodesk (*SW Utilities, CAD*)
111 McInnis Parkway
San Rafael, CA 94903
T: 415-507-5000
FaxBack: 415-507-5595
CIS: go ACAD
CIS: go AMMEDIA
FTP: ftp://ftp.autodesk.com/
URL: http://www.autodesk.com/
(Includes indexes for worldwide references)

Autotime Corporation (*Peripherals, Memory*)
6605 SW Macadam
Portland, OR 97201
T: 503-452-8577
F: 503-452-8495
FaxBack: 503-452-0208
E-mail: info@autotime.com
URL: http://www.teleport.com/~autotime/

Avance Logic (*ICs, Logic*)
T: 510-226-9555
E-mail: sales@avance.com
URL: http://www.avance.com/

AVerMedia (*Equipment, Video*)
47923A Warm Springs Blvd.
Fremont, CA 94539
T: 510-770-9899
F: 510-770-9901
URL: http://www.aver.com/aver/

Avex Electronics, Inc.
(*Manufacturer, Custom*)
URL: http://www.huber.com/Avex/avex95/avexmain.htm
(Includes links to worldwide facilities)

Avnet (*Computers, Systems*)
URL: http://www.avnet.com/
(Includes references to worldwide offices)

Avtech Electrosystems, Ltd.
(*Equipment, Test*)
PO Box 265
Ogdensburg, NY 13669-0265
T: 315-472-5270
F: 613-226-5772
E-mail: info@avtechpulse.com
URL: http://www.avtechpulse.com/

Award Software International
(*ICs, BIOS*)
777 E. Middlefield Rd.
Mountain View, CA 94043
T: 415-968-4433
F: 415-968-0274
B: 408-370-3139
URL: http://www.award.com/

Axis Communications (*Networking, CD-ROM and Printing*)
4 Constitution Way
Woburn, MA 01801-9755
T: 617-938-1188
F: 617-938-6161
E-mail: info@axisinc.com
URL: http://www.axis.se/

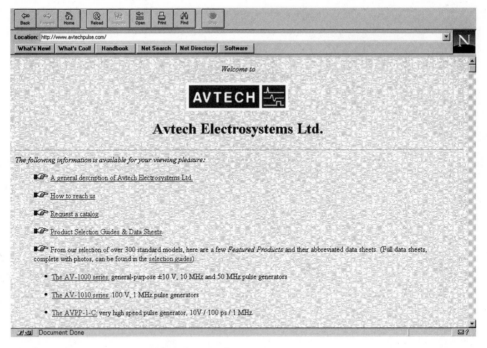

Figure 3-11 Avtech Electrosystems, Ltd.

Aztech Labs, Inc. (*Peripherals, Sound Boards*)
47811 Warm Springs Blvd.
Fremont, CA 94539
T: 800-258-2088
F: 510-623-8989
FTP:
ftp://ftp.aimnet.com/pub/users/aztech
URL: http://www.aztechca.com/

Banyan Systems (*Networking, Software*)
120 Flanders Rd.
Westboro, MA 01581
T: 508-898-1000
F: 508-898-1755
URL: http://www.banyan.com/

Bapco (*Diagnostics, Benchmarks*)
T: (408)988-7654
URL: http://www.bapco.com/
(List of benchmarking information and software)

B&B Electronics Manufacturing Co.
(*Automation, Equipment*)
PO Box 1040, Ottawa, IL 61350
T: 815-433-5100
F: 815-434-7094
E-mail: techsupt@bb-elec.com
(Technical Support)
E-mail: catrqst@bb-elec.com
(Catalog Request)
URL: http://www.bb-elec.com/

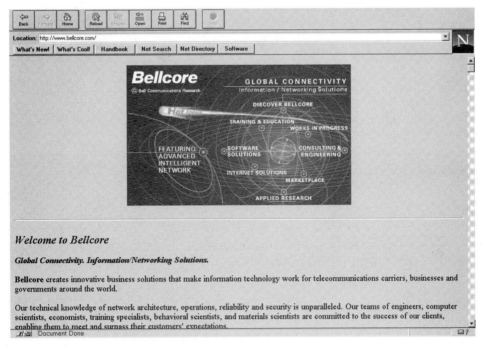

Figure 3-12   Bellcore.

Bay Networks   (*Networking, Hardware*)
4401 Great American Parkway
Santa Clara, CA 95054
T: 800-231-4213
CIS: go BAYNET
E-mail: answers@baynetworks.com
E-mail: baynetlit@selectnet.com
URL: http://www.baynetworks.com/

Beame and Whiteside Software
(*SW Utilities, Internet*)
Hummingbird Communications Ltd.
1 Sparks Ave.
North York, Ontario, M2H 2W1 Canada
T: 416-496-2200
F: 416-496-2207
B: 416-496-9233
E-mail: sales@hummingbird.com

E-mail: support@hummingbird.com
FTP: ftp://ftp.hummingbird.com/
URL: http://www.hummingbird.com/
bwsletter.html

Bellcore   (*Networking, Hardware*)
445 South St.
Morristown, NJ 07960-6438
T: 201-829-2000
URL: http://www.bellcore.com/

Bell Industries   (*Distributor,*
Components)
11812 San Vicente Blvd., Suite 300
Los Angeles, CA 90049
T: 800-525-6666
F: 800-444-0139
URL: http://www.bellind.com/
(Includes links to worldwide offices)

Benchmarq Microelectronics
(*ICs, Power Management*)
17919 Waterview Pkwy.
Dallas, TX 75252
T: 214-437-9195
F: 214-437-9198
URL: http://synapse-group.com/
Benchmarq/
URL: http://www.benchmarq.com/
(Includes links to worldwide offices)

Berg Electronics (*Computers, Connectors*)
150 Corporate Center., #201
Camp Hill, PA 17011
T: 800-237-2374
F: 717-938-7604
URL: http://www.bergelect.com/
(Includes links to worldwide offices)

Berkeley Software Design (*Computers, Servers*)
5575 Tech Center Dr., #110
Colorado Springs, CO 80918
T: 719-593-9445
F: 719-598-4238
Email: bsdi-info@bsdi.com
URL: http://www.bsdi.com/

Betacorp Multimedia (*Publisher, Multimedia*)
6770 Davand Dr. Ste. 40
Mississauga, Ontario, L5T 2G3 Canada
T: 905-564-2424
F: 905-564-6655
B: 905-564-8245
CIS: go BETACORP
E-mail: support@betacorp.com
FTP: ftp://ftp.betacorp.com
URL: http://www.betacorp.com/

Bitstream, Inc. (*SW Utilities, Fonts*)
215 First St.
Cambridge, MA 02142
T: 617-497-6222
F: 617-868-0784
E-mail address: info@bitstream.com
URL: http://www.bitstream.com/

Black Box Corporation (*Networking, Hardware*)
PO Box 12800
Pittsburgh, PA 15241
T: 412-746-5500
F: 412-746-0746
URL: http://www.blackbox.com/

Blue Sky Software Corporation
(*SW Utilities, Business*)
7777 Fay Ave., Suite 201
La Jolla, CA 92037
T: 619-459-6365
F: 619-459-6366
B: 619.551.2495
CIS: 73473,3636
Email: info@blue-sky.com
URL: http://www.blue-sky.com/

Boca Research, Inc. (*Computers, Modems Controllers, Video*)
1377 Clint Moore Rd.
Boca Raton, FL 33487-2722
T: 407-997-6227
F: 407-994-5848
B: 407-241-1601
URL: http://www.bocaresearch.com/

Borland International (*SW Utilities, Database/SW Utilities, Development*)
100 Borland Way
Scotts Valley, CA 95066
Phone: +1 408-431-1000
T: 800-841-8180 (Support Directory)
B: 408-431-5096

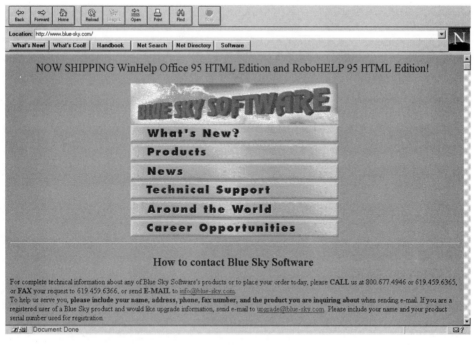

Figure 3-13   Blue Sky Software Corporation.

FaxBack: 800-822-4296
CIS: go BORLAND
URL: http://www.borland.com/

B-Plan   (*SW Utilities, Business*)
27 Maskit St.
Herzelia 46733, Israel
T: 972 9 562002
F: 972 9 574055
E-mail: bplan@elronet.co.il
URL:
http://www.netvision.net.il/~bplan/

Bristol Technology   (*SW Utilities,
Development*)
241 Ethan Allen Highway
Ridgefield, CT 06877
T: 203-438-6969
F: 203-438-5013

E-mail: info@bristol.com
URL: http://www.bristol.com/

Bromley Instruments   (*Manufacturer,
Custom*)
Ashton House
Farnborough Hill
Kent, BR6 6DA UK
T: +44 (0)1959 571085
URL: http://www.cityscape.co.uk/users/
ft64/index.html

Brooktree Corporation   (*ICs, Analog*)
9868 Scranton Rd.
San Diego, CA 92121
URL: http://www.brooktree.com/
(Includes links to worldwide distributors)

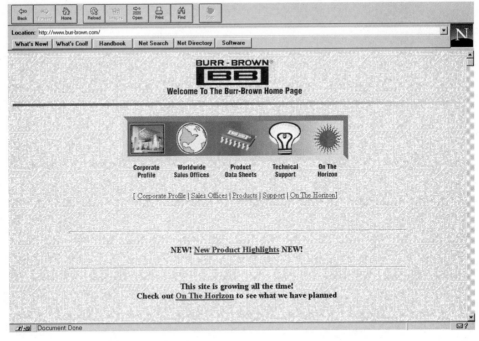

Figure 3-14 Burr-Brown.

Brother International Corporation
(*Computers, Printers*)
200 Cottontail Ln.
Sommerset, NJ 08875
T: 908-356-8880
F: 908-764-4481
B: 714-859-2610
URL: http://www.brother.com/

Burr-Brown (*ICs, Analog*)
PO Box 11400
Tucson, AZ 85734-1400
T: 520-746-1111
F: 520-746-7401
URL: http://www.burr-brown.com/

BusLogic (*Controllers, SCSI*)
4151 Burton Dr.
Santa Clara, CA 95054-1564
T: 408-492-9090

F: 408-492-1542
B: 408-492-1984
URL: http://www.buslogic.com/

Butterworth Heinemann (*Publisher,
Electronics*)
313 Washington St.
Newton, MA 02158-1626
T: 617-928-2500
F: 617-928-2620
URL: http://www.bh.com/bh/

Cabletron Systems, Inc. (*Networking,
Software*)
36 Industrial Way
Rochester, NH 03867
T: 603-332-9400
E-mail: support@ctron.com
URL: http://www.ctron.com/

Cadence (*Simulation, Circuit CAD*)
555 River Oaks Parkway
San Jose, CA 95134
T: 408-943-1234
F: 408-943-0513
URL: http://www.cadence.com/

CAD-Migos Software Tools, Inc.
(*Simulation, Circuit CAD*)
499 Seaport Ct., Suite 201
Redwood City, CA 94063
T: 415-369-5853
URL: http://www.cadmigos.com/

Caere Corporation (*SW Utilities, Imaging*)
100 Cooper Ct.
Los Gatos, CA 95030
T: 408-395-7000
URL: http://www.caere.com/

California Switch & Signal (Cal Switch) (*Manufacturer, Components*)
1010 Sandhill Ave.
Carson, CA 90746-1373
T: 310-632-4300
F: 310-632-4301
URL: http://www.calswitch.com/

Caligari (*SW Utilities, CAD*)
1935 Landings Dr.
Mountain View, CA 94043
T: 415-390-9600
F: 415-390-9755
E-mail: support@caligari.com
URL: http://www.caligari.com/

Canon USA, Inc. (*Computers, Printers*)
One Canon Plaza
Lake Success, NY 11042
T: 516-488-6700
F: 516-354-5805
B: 516-488-6528

URL: http://www.canon.com/
URL: http://www.usa.canon.com/

Capilano Computing Systems
(*Simulation, Circuit CAD*)
T: 604-522-6200
E-mail: info@capilano.com
URL: http://www.capilano.com/

Capsoft Development Corporation
(*SW Utilities, Development*)
732 East Utah Valley Dr., Suite 400
American Fork, UT 84003
T: 801-763-3900
F: 801-763-3999
B: 801-763-9814
URL: http://www.capsoft.com/

Cardinal Technologies (*Controllers, Video/Computers, Modems*)
T: 717-293-3000
AOL: keyword CARDINAL
E-mail: techs@cardtech.com

Cascade (*Simulation, Circuit CAD*)
3650 131st Ave., S.E., Suite 650
Bellevue, WA 98006
T: 206-643-0200
F: 206-649-7600
E-mail: info@cdac.com
URL: http://www.cdac.com/

CAST (*Services, Education*)
24 White Birch Dr.
Pomona, NY 10970
T: 914-354-4945
F: 914-354-0325
E-mail: info@cast-inc.com
URL: http://www.cast-inc.com/

Figure 3-15 Canon USA, Inc.

CDW Computer Centers, Inc.
(*Distributor, PC Assemblies*)
1020 E. Lake Cook Rd.
Buffalo Grove, IL 60089
T: 800-726-4239
F: 847-465-6800
URL: http://www.cdw.com/

Central Data (*Controllers, I/O*)
T: 217-359-8010
F: 217-359-6904
E-mail: support@cd.com
FTP: ftp.cd.com
URL: http://www.cd.com/

Central Point Software, Inc.
(*Diagnostics, Software*)
15220 NW Greenbrier Parkway
Beaverton, OR 97006-9937

T: 503-690-8088
F: 503-690-8083
B: 503-690-6650

Cera Research (*Publisher, Computers*)
36608 Newark Blvd.
Newark, CA 94560
T: 510-745-7928
F: 510-745-7929
E-mail: inquiry@eg3.com
URL: http://www.cera.com/index.html

Cermetek Microelectronics (*Manufacturer, Components*)
T: 408-752-5000
E-mail: cermetek@ix.netcom.com
URL: http://www.isi.net/ebs/cermetek/index.html

Figure 3-16   CE Software.

CE Software   (*SW Utilities, Internet*)
P.O. Box 65580
1801 Industrial Circle
West Des Moines, IA 50265
T: 515-221-1801
F: 515-221-1806
Tech. Supp: 515-221-1803
Tech. Fax: 515-221-2169
QuickFacts: 515-221-2168
B: 515-221-2167
URL: http://www.cesoft.com/

Cheyenne Software   (*Networking, Software*)
3 Expressway Plaza
Roslyn Heights, NY 11577
T: 516-465-4000
F: 516-484-2489
Tech. Supp: 800-243-9832

Tech. Fax: 516-465-5115
B: 516-465-3900
FaxBack: 516-465-5979
InocuLAN GetBBS: (516) 465-3930
CompuServe: go CHEYENNE
FTP: ftp://ftp.cheyenne.com/
URL: http://www.chey.com/
URL: http://www.cheyenne.com

Chinon America   (*Computers, Drives*)
660 Maple Ave.
Torrance, CA 90503
T: 310-533-0274
F: 310-533-1727
B: 310-320-4160

Chip Express   (*Manufacturer, ICs*)
2903 Bunker Hill Lane, Suite 105
Santa Clara, CA 95054
T: 408-988-2445
F: 408-988-2449
URL: http://www.chipexpress.com/

Chips & Technologies   (*ICs, Logic*)
4950 Zanker Rd.
San Jose, CA 95134
T: 408-434-0600
F: 408-894-2079
B: 408-456-0721
URL: http://www.chips.com/

Chromatic Research   (*ICs, Logic*)
URL: http://www.mpact.com/

Chrontel   (*ICs, Logic*)
2210 O'Toole Ave.
San Jose, CA 95131-1326
T: 408-383-9328
F: 408-383-9338
E-mail: products@chrontel.com
URL: http://www.docwriter.com/chron-tel/

Chrysalis Symbolic Design, Inc.
(*Simulation, Circuit CAD*)
101 Billerica Ave., Bldg. 5
North Billerica, MA 01862
T: 508-436-9909
F: 508-436-9697
E-mail: silver@chrysalis.com
URL: http://www.chrysalis.com/

Circuit Cellar, Inc.   (*Publisher, Electronics*)
4 Park St.
Vernon, CT 06066
T: 860-875-2751
F: 860-872-2204

E-mail: sales@circellar.com
URL: http://tfnetils.unc.edu/cgi-bin/CCBBS/
URL: http://www.circellar.com/ink.html

Circuit Specialists, Inc.   (*Distributor, Components*)
220 South Country Club Dr., Bldg. 2
Mesa, AZ 85210-1248
T: 800-811-5208
F: 602-464-5824
URL: http://www.cir.com/

Circuit Works   (*Manufacturer, PCB*)
85 W. Sylvania Ave.
Neptune, NJ 07753
T: 908-774-1811
F: 908-774-7198
E-mail: ETCHANT@IX.NETCOM.COM
URL: http://www.localweb.com/circuitworks/test1.html

Cirrus Logic   (*ICs, Logic*)
3100 W. Warren Ave.
Fremont, CA 94538
T: 510-623-8300
F: 510-226-2270
FTP: ftp://ftp.cirrus.com/
URL: http://www.cirrus.com/

Cisco   (*Networking, Hardware*)
170 West Tasman Dr.
San Jose, CA 95134-1706
Tel: 408-526-4000
F: 408-526-4100
URL: http://www.cisco.com/

Citizen America Corporation
(*Computers, Printers*)
2450 Broadway, Suite 600
Santa Monica, CA 90404

Location: http://www.cir.com/

Welcome To Our Online Catalog

CIRCUIT SPECIALISTS INC.

Choose Your Selection:

DAC/PLC
(Data Acquisition & Control/Programmable Logic Controllers)

Industrial Computers & Peripherals

Document Done

Figure 3-17   Circuit Specialists, Inc.

T: 310-453-0614
F: 310-453-2814
B: 310-453-7564

C. Itoh   (*Computers, Printers*)
2701 Dow Ave.
Tustin, CA 92680
T: 800-877-14221
F: 714-757-4488
B: 714-573-2645

Claris Corporation   (*SW Utilities, Business*)
T: 408-727-9004
FaxBack: 408-987-3900
AOL: keyword CLARIS
CIS: go CLARIS
URL: http://www.claris.com/

CMO   (*Distributor, PC Assemblies*)
101 Reighard Ave.
Williamsport, PA 17701
T: 717-327-9200
F: 717-327-1217
URL: http://www.cmo.newmmi.com/

Cogent Data   (*Networking, Hardware & Software*)
640 Mullis
Friday Harbor, WA 98250
T: 360-378-2929
F: 360-378-2882
URL: http://www.cogentdata.com/

Collins Printed Circuits (Rockwell)
(*Manufacturer, PCB*)
400 Collins Rd. NE
Cedar Rapids, IA 52498
URL: http://www.rockwell.com/
rockwell/bus_units/cca/cpc/

Colorado Memory Systems, Inc.
(*Peripherals, Tape*)
800 S. Taft Ave.
Loveland, CO 80537
T: 303-669-8000
F: 303-667-0997
B: 303-635-0650
URL: http://www.hp.com/cms/html/
products.htm

Columbia Data Products (*Peripherals,
Tape/Services, Data Recovery*)
1070-B Rainer Dr.
PO Box 142584
Altamonte Springs, FL 32714
T: 407-869-6700
F: 407-862-4725
B: 407-862-4724
E-mail: cdpi@cdpi.com
URL: http://www.cdp.com/
(Home page)
URL: http://www.cdp.com/so/
(Storage Online Newsletter)

ComCom Systems (*Equipment,
Imaging*)
2420 Enterprise Rd., Suite 201
Clearwater, FL 34623
T: 813-725-3200
F: 813-796-6596
E-mail: support@comcomsystems.com
URL: http://www.comcomsystems.com/
image.html

Compaq Computer Corporation
(*Computers, Systems*)
20555 SH 249
Houston, TX 77070
T: 713-370-0670
CIS: go COMPAQ
E-mail: support@compaq.com
URL: http://www.compaq.com/

Component Distributors, Inc.
(*Distributor, Components*)
T: 800-777-7334
URL: http://www.compdist.com/
(Includes references for nation-wide con-
tacts)

Compression Labs, Inc. (*Controllers,
Video*)
2860 Junction Ave.
San Jose, CA 95134-1900
T: 408-435-3000
F: 408-922-5429
URL: http://www.clix.com/

Comp-U-Plus (*Distributor, PC
Assemblies*)
20 Robert Pitt Dr.
Monsey, NY 10952
T: 914-352-8100
URL: http://www.compuplus.com/

CompUSA Direct (*Distributor, PC
Assemblies*)
T: 214-888-5770
F: 214-888-5706
URL: http://www.compusa.com/

CompuServe (*Services, Online*)
P.O. Box 20212
Columbus, OH 43220
URL: http://www.compuserve.com/

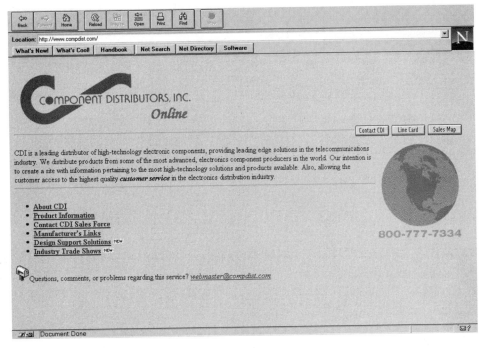

Figure 3-18   Component Distributors, Inc.

ComputAbility (*Distributor, PC Assemblies*)
P.O. Box 17882
Milwaukee, WI 53217
T: 800-554-9950
F: 414-357-7814
URL: http://www.computability.com/

Computer Associates (*SW Utilities, Business*)
One Computer Associates Plaza
Islandia, NY 11788-7000
T: 516-342-5224
CIS: go CAI
URL: http://www.cai.com/

Computer Connections (*Publisher, Computers*)
8499 S. Tamiami Tr., Ste 6
Sarasota, FL 34238
T: 941-966-1374
F: 941-966-2854
E-mail: digigraf@digigraf.com

Computer Currents (*Publisher, Computers*)
5720 Hollis St.
Emeryville, CA 94608
T: 510-547-6800
F: 510-547-4613
CIS: 73554,3010
URL: http://www.ccurrents.com/

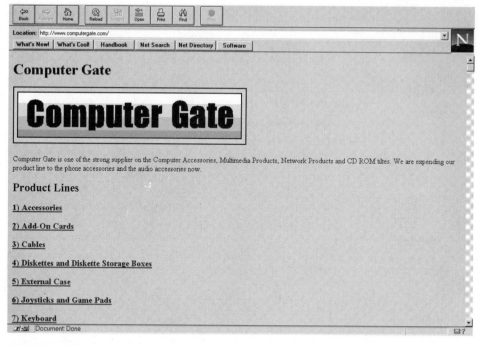

Figure 3-19 Computer Gate.

Computer Discount Warehouse (CDW)
(*Computers, Systems*)
1020 E. Lake Cook Rd.
Buffalo Grove, IL 60089
T: 708-465-6000
F: 708-465-6800
B: 708-465-6899

Computer Friends (*Recycling, Fill Kits*)
14250 NW Science Park Dr.
Portland, OR 97229
T: 503-626-2291
F: 503-643-5379
E-mail: cfriends@teleport.com
URL:
http://www.teleport.com/~cfriends/

Computer Gate International
(*Distributor, Supplies*)
2960 Gordon Ave.
Santa Clara, CA 95051
T: 408-730-0673
F: 408-730-0735
E-mail: gate@aimnet.com
URL: http://www.computergate.com/

Computer Mail Services (*Services, Online*)
T: 810-352-8385
B: 810-352-8386
E-mail: 259-6798@mcimail.com

Computer Marketplace, Inc.
(*Distributor, PC Assemblies*)
1101 Sussex Blvd.
Broomall, PA 19008

T: 610-690-6900
F: 610-690-6909
Email: sales@cmpexpress.com
URL:
http://www.cmpexpress.com/cmp/

Computer Parts Unlimited
(*Distributor, Components*)
5069 Maureen Ln.
Moorpark, CA 93021
T: 805-523-2500
F: 805-532-2599
E-mail: comppart@aol.com

Computer Network Technology Corp.
(*Networking, Hardware*)
6500 Wedgwood Rd.
Maple Grove, MN 55311-3640
T: 612-550-8000
F: 612-550-8800
URL: http://www.cnt.com/

Computer Shopper (*Publisher, Computers*)
URL: http://www.zdnet.com/cshopper/

Connect, Inc. (*SW Utilities, Internet*)
515 Ellis St.
Mountain View, CA 94043-2242
T: 415-254-4000
F: 415-254-4800
E-mail: info@connectinc.com
URL: http://www.connectinc.com/

Connectix Corporation (*Equipment, Video*)
2655 Campus Dr.
San Mateo, CA 94403
T: 415-571-5100
F: 415-571-5195
E-mail: info@connectix.com/

ConnectSoft (*SW Utilities, Business*)
T: 206-803-8385
CIS: 71333,2244
E-mail: techsupp@connectsoft.com

Connectware, Inc. (*Networking, Software*)
1301 E. Arapaho Rd.
Richardson, TX 75081
T: 214-907-1093
F: 214-907-1594
URL: http://www.rtp.connectware.com/

Conner Peripherals, Inc. (*Computers, Drives*)
3081 Zanker Rd.
San Jose, CA 95134
T: 408-456-4500
F: 407-262-4225
FaxBack: 408-456-4903 (West)
FaxBack: 407-262-4755 (East)
B: 407-263-3502
URL: http://www.conner.com/
(Merging with Seagate)

Continental Resources (*Services, Repair*)
175 Middlesex Turnpike
Bedford, MA 01730
T: 800-937-4688
URL: http://www.conres.com/

Convex (*Computers, Systems*)
3000 Waterview Parkway
P.O. Box 833851
Richardson, TX 75083-3851
T: 214-497-4000
F: 214-497-4848
Gopher: gopher://gopher.convex.com/
URL: http://www.convex.com/

Figure 3-20 Corcom.

Copper Electronics (*Equipment, Communication*)
T: 800-626-6343
URL:
http://www.iglou.com/copper/index.cgi

Corcom (*Manufacturer, Components*)
844 E. Rockland Rd.
Libertyville, IL 60048
T: 847-680-7400
F: 847-680-8169
Email: info@cor.com
URL: http://www.cor.com/home.html

Core International, Inc. (*Distributor, Drives*)
7171 North Federal Hwy.
Boca Raton, FL 33487

T: 407-997-6055
B: 407-241-2929

Corel Corporation (*SW Utilities, DTP*)
1600 Carling Ave.
Ottawa, ON, K1Z 8R7 Canada
T: 613-728-8200
F: 613-761-9176
URL: http://www.corel.ca/

Corning (*Manufacturer, Materials*)
PO Box 7429, Union Station
Endicott, NY 13760
T: 800-525-2524
F: 800-834-3504
E-mail: fiber@corning.com
URL: http://usa.net/corning-fiber/

Corporate Disk (*Manufacturing, Software*)
1226 Michael Dr.
Wood Dale, IL 60191
T: 800-634-3475
URL: http://www.disk.com/

Corporate Systems Center
(*Distributor, Drives*)
500 Lawrence Expressway, Unit H
Sunnyvale, CA 94086
T: (408) 734-DISK
URL:
http://www.corpsys.com/index.html

Cray Research, Inc. (*Computers, Supercomputers*)
655 Lone Oak Dr.
Eagan, MN 55121
T: 612-452-6650
F: 612-683-7199
URL: http://www.cray.com/

Creative Labs (*Controller, Video/Peripherals, Sound*)
Technical Support
1523 Cimarron Plaza
Stillwater, OK 74075
T: 405-742-6622
F: 405-742-6633
B: 405-742-6660 (14.4KB)
CIS: go BLASTER
URL: http://www.creaf.com/

Criterion Software (*SW Utilities, Development*)
Westbury Ct.
Buryfields
Guildford, Surrey, GU2 5AZ UK
T: +44-1483-406200
F: +44-1483-406211
E-mail: rw-info@criterion.canon.co.uk

URL: http://www.criterion.canon.co.uk/csl/cslhome.html
URL: http://www.canon.co.uk/csl/

CrossComm Corporation
(*Networking, Hardware*)
450 Donald Lynch Blvd.
Marlborough, MA 01752
T: 508-481-4060
F: 508-229-5535
E-mail: marcomm@crosscomm.com
URL: http://www.crosscomm.com/

CSI (Colorado Superconductor Inc.)
(*Manufacturer, Materials*)
P.O. Box 8223
Fort Collins, CO 80524
T: 970-490-2787
F: 970-490-1301
URL: http://www2.csn.net/~donsher/

CST (*Equipment, Test*)
2336 Lu Field Rd.
Dallas, TX 75229
T: 214-241-2662
F: 214-241-2661
B: 214-241-3782

CTS (*Manufacturer, Components*)
406 Parr Rd.
Berne, IN 46711
T: 219-589-3111
F: 219-589-3243
URL: http://www.fwi.com/cts/

CyberMax Computer, Inc.
(*Computers, Systems*)
133 N. 5th St.
Allentown, PA 18102
T: 610-770-1808
URL: http://www.cybmax.com/

Figure 3-21 CSI.

Cybex Corporation (*Equipment, Communication*)
4912 Research Dr.
Huntsville, AL 35805
T: 205-430-4000
F: 205-430-4030
FaxBack: 800-GO-CYBEX
URL: http://www.cybex.com/

Cypress Semiconductor (*ICs, Logic*)
3901 N. First St.
San Jose, CA 95134
T: 408-943-2600
URL: http://www.cypress.com/

Cyrix (*ICs, Microprocessors*)
2703 N. Central Expressway
Richardson, TX 75080
T: 214-994-8388

F: 214-699-9857
E-mail: tech_support@cyrix.com
URL: http://www.cyrix.com/

DAKCO (*Computers, Systems*)
9600 Perry Highway
Pittsburgh, PA 15237
T: 412-369-8400
F: 412-369-0940
E-mail: dakco@telerama.lm.com
URL: http://dakco.lm.com/

Dallas Semiconductor (*ICs, Logic*)
4401 South Beltwood Parkway
Dallas, TX 75244
T: 214-450-0400
F: 214-450-3715
FTP: ftp://ftp.dalsemi.com/
URL: http://www.dalsemi.com/

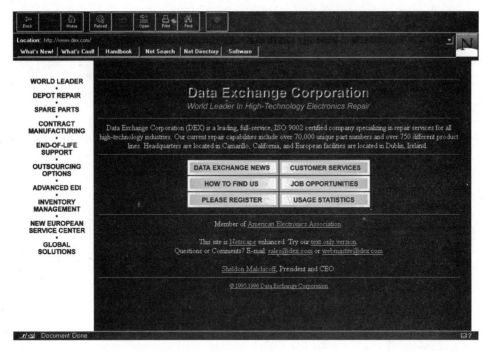

Figure 3-22 Data Exchange Corporation.

Data Comm Warehouse (*Distributor, PC Assemblies and Software*)
1720 Oak St.
P.O. Box 301
Lakewood, NJ 08701-9885
T: 800-328-2261
F: 908-363-4823
FaxBack: 203-854-5300
CIS: go MCW
URL: http://www.warehouse.com/

Data Depot, Inc. (*Diagnostics, Hardware and Software*)
1710 Drew St.
Clearwater, FL 34615
T: 813-446-3402
F: 813-443-4377
E-mail: info@datadepo.com
URL: http://www.datadepo.com/

Data Exchange Corporation (*Services, Repair*)
3600 Via Pescador
Camarillo, CA 93012
T: 805-388-1711
F: 805-482-4856
E-mail: sales@dex.com
URL: http://www.dex.com/

Data General Corporation (*Computers, Systems*)
4400 Computer Dr.
Westboro, MA 01580
T: 508-898-5071
URL: http://www.dg.com/

Data I/O   (*Simulation, Circuit CAD*)
10525 Willows Rd. NE
PO Box 97046
Redmond, WA 98073-9746
T: 206-881-6444
F: 206-882-1043
URL: http://www.data-io.com/

Dataproducts, Inc.   (*Computers, Printers*)
1757 Tapo Canyon Rd., Suite 200
Simi Valley, CA 93063-3393
T: 805-578-4000
F: 805-578-4001
B: 805-578-9251 (14.4KB)
FaxBack: 805-578-9255
E-mail: info@dpc.com
URL: http://www.dpc.com/

Datastorm Technologies   (*SW Utilities, Communication*)
3212 Lemone Blvd.
Columbia, MO 65205
T: 314-875-0530
F: 314-499-1552
FaxBack: 314-499-4556
B: 314-875-0503
CIS: go DATASTORM

Data-Tech Institute   (*Services, Education*)
PO Box 2429
Clifton, NJ 07015
T: 201-478-5400
URL: http://www.datatech.com/

Data Technology Corporation (DTC)
(*Distributor, PC Assemblies*)
1515 Centre Pointe Dr.
Milpitas, CA 95035-8010
T: 408-942-4000
F: 408-942-4027
B: 408-942-4197

Daybreak Communications
(*Distributor, Components*)
T: 800-440-7278
E-mail: info@daybreak.com
URL: http://www.daybreak.com/

Daystar Digital   (*Computers, Systems*)
5556 Atlanta Highway
Flowery Branch, GA 30542
T: 770-967-2077
F: 770-967-3018
E-mail: info@daystar.com
E-mail: support@daystar.com
URL: http://www.daystar.com/

DC Drives   (*Distributor, Drives*)
3716 Timber Dr.
Dickinson, TX 77539
T: 713-534-4140
F: 713-534-6452
URL: http://www.hard-drive.com/

Definitive Technology   (*Peripherals, Speakers*)
11105 Valley Heights Dr.
Baltimore, MD 21117
T: 410-363-7148
E-mail: definitive@soundsite.com
URL:
http://www.soundsite.com/index.html

Delco Electronics   (*Manufacturer, ICs*)
501 Ang Mo Kio
Industrial Park 1
Singapore 569621
URL: http://www.singapore.com/
companies/delco/

Dell Computer Corporation
(*Computers, Systems*)
9505 Arboretum Blvd.
Austin, TX 78759
T: 512-338-4400

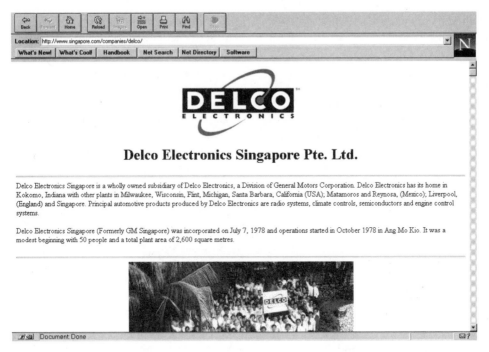

Figure 3-23 Delco Electronics.

F: 512-728-3653
B: 512-338-8528
URL: http://www.us.dell.com/

Delorme (*SW Utilities, Business*)
Lower Main St.
P.O. Box 298
Freeport, ME 04032
T: 800-452-5931
F: 800-575-2244
FaxBack:207-865-7083
URL: http://www.delorme.com/

Delrina Corporation (*SW Utilities,*
Communication)
T: 416-441-1604
FaxBack: 416-443-1614
B: 416-441-2752
AOL: keyword DELRINA
CIS: go DELRINA

E-mail: winfax.support@delrina.com
E-mail: wincomm.support@delrina.com
URL: http://www.delrina.com/

Deneba Software (*SW Utilities,*
Business)
AOL: keyword DENEBA
CIS: 76004,2154

Develcon (*Networking, Hardware*)
856 51st St. East
Saskatoon, SK, S7K 5C7 Canada
T: 306-933-3300
F: 306-931-1370
URL: http://www.develcon.com/

Dialogic   (*Equipment, Communication*)
1515 Route Ten
Parsippany, NJ 07054
T: 201-993-3030
F: 201-993-3093
E-mail: sales@dialogic.com
URL: http://www.dialogic.com/

Diamond Multimedia Systems, Inc.
(*Controllers, Video*)
2880 Junction Ave.
San Jose, CA 95134-1922
T: 408-325-7000
F: 408-325-7070
B: 408-325-7080 (to 14.4KB)
B: 408-325-7175 (to 28.8KB)
AOL: keyword DIAMOND
CIS: go DIAMOND
FTP: ftp.diamondmm.com
URL: http://www.diamondmm.com/

DigiBoard   (*Networking, Hardware*)
6400 Flying Cloud Dr.
Eden Prairie, MN 55344
T: 612-943-9020
F: 612-943-5398
E-mail: sales@digibd.com
URL: http://www.digibd.com/

Digi-Key Corporation   (*Distributor,*
Components)
701 Brooks Ave. S.
PO Box 677
Thief River Falls, MN 56701-0677
F: 218-681-3380
URL: http://www.digikey.com/

Digital Equipment Corporation
(*Computers, Systems*)
111 Powdermill Rd.
Maynard, MA 01754

T: 508-493-5111
URL: http://www.dec.com/info.html

Digital Vision, Inc.   (*Equipment, Video*)
270 Bridge St.
Dedham, MA 02026
T: 617-329-5400
URL: http://www.digvis.com/digvis/

Direct Silicon Design
(*Manufacturer, ICs*)
T: 714-553-0944
E-mail: info@dsdi.com
URL: http://www.dsdi.com/designs/

Direct Wave   (*Computers, Systems*)
4260 East Brickell
Ontario, CA 91761
T: 909-390-8058
F: 909-390-8061
B: 909-390-8052
E-mail: drctwave@ix.netcom.com
URL: http://www.directwave.com/

Disaster Recovery Journal (DRJ)
(*Publisher, Computers*)
P.O. Box 510110
St. Louis, MO 63151
T: 314-894-0276
F: 314-894-7474
E-mail: drj@mo.net
URL: http://www.drj.com

Display Tech Multimedia   (*Equipment,*
Video)
E-mail: dtmi@ccnet.com
URL: http://www.ccnet.com/~dtmi/

Distributed Processing Technology
(DPT)   (*Controllers, SCSI*)
140 Candace Dr.
Maitland, FL 32751

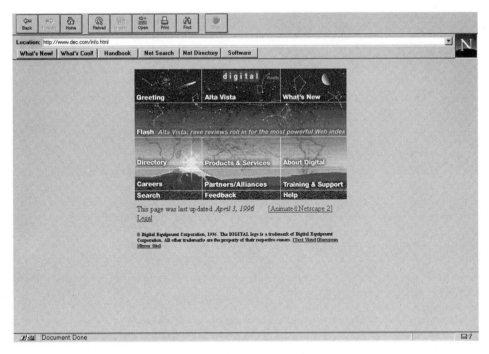

Figure 3-24 Digital Equipment Corporation.

T: 407-830-5522
F: 407-260-5366
B: 407-831-6432
E-mail: techsupp@dpt.com

Ditek Software Corporation
(*SW Utilities, Business*)
2800 John St., #15
Markham, Ontario, L3R 0E2 Canada
T: 905-479-1990
CIS: 74774,2225
URL: http://www.ditek.com/

D-Link Systems, Inc. (*Networking,*
Hardware)
T: 714-455-1688
B: 714-455-9616
E-mail: tech@irvine.dlink.com
URL: http://www.dlink.com/

Dolby Laboratories (*Peripherals,*
Sound)
100 Potrero Ave.
San Francisco, CA 94103-4813
URL:
http://www.dolby.com/ht/ds&pl/
ds&dspl.html

Dove Systems (*Equipment, Industrial*)
3563 Sueldo St. Unit E
San Luis Obispo, CA 93401
T: 805-541-8292
F: 805-541-8293
URL:
http://www.dovesystems.com/dove/

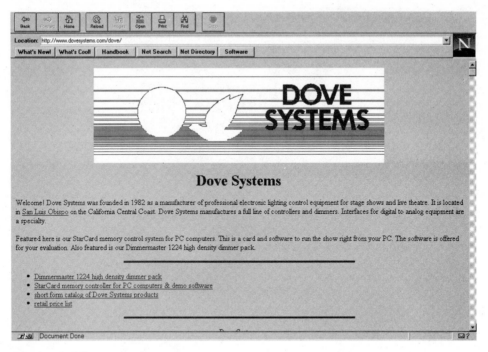

Figure 3-25 Dove Systems.

Doxa Audio Development
(*Equipment, Sound*)
Postboks 5017
N-4626 Kristiansand S, NORWAY
T: +47 38 01 60 19
F: +47 38 01 63 44
E-mail: lauvland@telepost.no
URL:
http://www.ifi.uio.no/~stiand/Doxa/
index.html

DSP Group (*Peripherals, Sound*)
3120 Scott Blvd.
Santa Clara, CA 95054-3317
T: 408-986-4300
F: 408-986-4323
URL: http://www.dspg.com/

DSPnet (*SW Utilities, Development*)
URL: http://www.dspnet.com/
(Includes a wealth of links)

D&S Technologies (*Equipment,
Security*)
202 Milton St.
Dedham, MA 02026
T: 617-461-0171
F: 617-461-0216
E-mail: dstech@tiac.net
URL:
http://www.tiac.net/users/dstech/html/

dTb Software (*Simulation, Circuit
CAD*)
Box 5057
Walnut Creek, CA 94596
T: 510-946-1043
F: 510-942-0690

E-mail: info@dtbsware.com
URL: http://www.dtbsware.com/

DTC (*Controllers, Drive*)
1515 Centre Pointe Dr.
Milpitas, CA 95035
T: 408-942-4000
F: 408-942-4027
URL: http://www.datatechnology.com/

DTI Factory Outlet (*Computers, Systems*)
South Braintree Park
1515 Washington St. (Route 37)
Braintree, MA 02184
T: 617-691-1200
F: 617-691-1300
E-mail: info@datatrend.com
URL: http://www.channel1.com/
business/datatrend/

DTK Computer, Inc. (*ICs, BIOS*)
770 Epperson Dr.
City of Industry, CA 91748
T: 818-810-0098
B: 818-333-6548

Dynamic Learning Systems (*Publisher, Computer Repair*)
P.O. Box 282
Jefferson, MA 01522-0282
T: 508-829-6744
F: 508-829-6819
B: 508-829-6706
CIS: 73652,3205
E-mail: sbigelow@cerfnet.com
URL: http://www.dlspubs.com/

Eagle Design Automation
(*Simulation, Circuit CAD*)
13515 SW Millikan Way
Beaverton, OR 97005
T: 503-520-2300

F: 503-520-2323
URL: http://www.eagledes.com/

Eastern Acoustic Works (*Peripherals, Sound*)
1 Main St.
Whitinsville MA 01588
T: 508-234-4295
F: 508-234-3776
E-mail: steved@eaw.com
URL: http://www.eaw.com/

Ecliptek (*Distributor, Components*)
3545 Cadillac Ave.
Costa Mesa, CA 92626-1401
T: 714-433-1200
F: 714-433-1234
FaxBack: 1-800-ECLIPTEK
E-mail: ecquality@ecliptek.com
URL:
http://www.ecliptek.com/ecliptek/

EDAC (*Computers, Connectors*)
URL: http://www.edac.net/
(Includes links to worldwide distributors)

EE Times (WAIS, Inc.) (*Publisher, Electronics*)
690 Fifth St.
San Francisco, CA 94107
T: 415-525-2553
Email: frontdesk@wais.com
URL: http://www.wais.com/eet/
current/

EJE Research (*Manufacturer, Custom*)
20 French Rd.
Buffalo, NY 14227
T: 716-668-6600
F: 716-656-9605
E-mail: ejemail@eje.com
URL: http://www.eje.com/

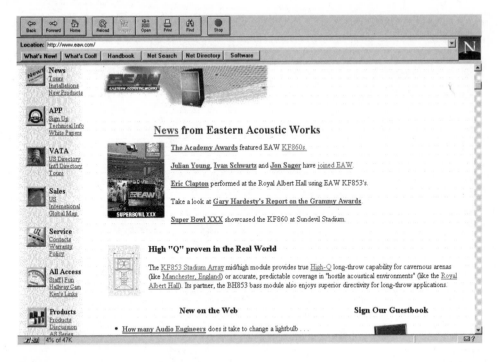

Figure 3-26   Eastern Acoustic Works.

ELANIX   (*Simulation, Circuit CAD*)
5655 Lindero Canyon Rd., Suite 721
Westlake Village, CA 91362
T: 818-597-1414
F: 818-597-1427
B: 818-597-0306
E-mail: elanix@elanix.com
E-mail: SystemView@elanix.com
URL: http://www.elanix.com/elanix/

ELAN Software Corporation
(*SW Utilities, Communication*)
T: 310-459-1222 (Including FaxBack
service)
F: 310-459-8222
B: 310-459-3443

Elantec   (*ICs, Logic*)
1996 Tarob Ct.
Milpitas, CA 95035
T: 408-945-1323
F: 408-945-9305
URL: http://www.elantec.com/

Elastic Reality   (*SW Utilities, Imaging*)
T: 508-640-3070
F: 508-640-9486
CIS: 71333,3020
E-mail: avidnewmed@aol.com

Electronics Marketing Corporation
(*Distributor, Components*)
1150 West Third Ave.
Columbus, OH 43212
T: 614-299-4161
F: 614-299-4121

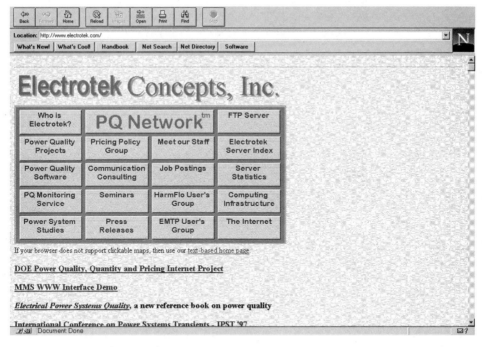

Figure 3-27   Electrotek Concepts, Inc.

E-mail: usemcrwm@ibmmail.com
URL:
http://www.smartpages.com/emc/

Electronix Corporation   (*Distributor,
Components*)
313 W. Main St.
Fairborn, OH 45324-5036
T: 513-878-1828
F: 513-878-1972
URL: http://www.electronix.com/
elexcorp/

Electroservice Laboratories   (*Services,
Repair*)
6085 Sikorsky St.
Ventura, CA 93003
T: 805-644-2944
F: 805-644-5006

B: 805-644-7810
E-mail: info@esl.com

Electrotek Concepts, Inc.   (*Equipment,
Power Products*)
408 North Cedar Bluff Rd., Suite 500
Knoxville, TN 37923-3605
T: 423-470-9222
F: 423-470-9223
URL: http://www.electrotek.com/

Elek-Tek, Inc.   (*Distributor,
PC Assemblies*)
7350 North Linder Ave.
Skokie, IL 60077
T: 708-677-7660
F: 708-677-1081
E-mail: info@elektek.com
URL: http://www.elektek.com/

**Embedded Support Tools Corp.
(ESTC)** (*Equipment, Emulation*)
120 Royall St.
Canton, MA 02021
T: 617-828-5588
F: 617-821-2268
E-mail: sales@estc.com
URL: http://www.estc.com/

EMD Associates (*Manufacturer,
Custom*)
T: 507-452-8932
F: 507-454-9028
URL: http://www.mps.org/emd/

Encore Computer Corporation
(*Computers, Systems*)
6901 West Sunrise Blvd.
Fort Lauderdale, FL 33313-4499
T: 305-587-2900
F: 305-797-5793
URL: http://www.encore.com/

Ensoniq (*Peripherals, Sound Boards*)
155 Great Valley Parkway
PO Box 3035
Malvern, PA 19355-0735
T: 610-647-3930
F: 610-647-8908
B: 610-647-3195
FaxBack: 800-257-1439
CIS: go MMENSONIQ
E-mail: multimedia@ensoniq.com
URL: http://www.ensoniq.com/

Enterprise Integration Technologies
(*SW Utilities, Internet*)
800 El Camino Real
Menlo Park, CA 94025-4808
T: 415-617-8000
F: 415-617-8019

E-mail: info@eit.com
URL: http://www.eit.com/

Environmental Laser (*Recycling, Toner
Cartridge*)
email: info@toners.com
URL: http://www.catalog.com/toner/
envlas.html

Envision Interactive Computers
(*Computers, Systems*)
130 N. Central Ave.
Glendale, CA 91203
T: 800-700-3119
F: 818-552-5992
URL: http://www.env.com/

Envisions (*Distributor, Scanners*)
47400 Seabridge Dr.
Fremont, CA 94538
T: 510-661-4357
F: 510-438-6709
URL: http://www.envisions.com/

Epson America, Inc. (*Computers,
Printers*)
20770 Madrona Ave.
Torrance, CA 90509-2842
T: 310-787-6300
F: 310-782-5350
B: 408-782-4531
URL: http://www.epson.com/

EPS Technologies, Inc. (*Computers,
Systems*)
10069 Dakota Ave.
Box 278
Jefferson, SD 57038
T: 605-966-5586
F: 605-966-5482
URL: http://www.epstech.com/

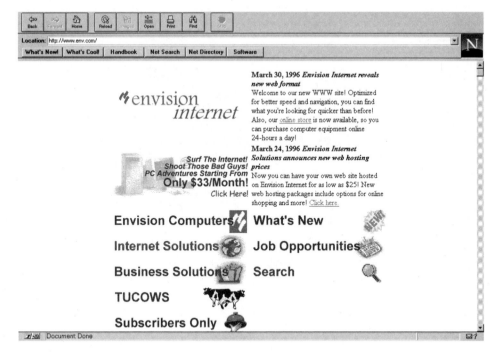

Figure 3-28   Envision Interactive Computers.

Ericsson   (*Equipment, Communication*)
700 East Dunne Ave., Suite 101
Morgan Hill, CA 95037
T: 408-778-9434
F: 408-779-3108
URL: http://www.ericsson.com/

Everex Systems   (*Computers, Systems*)
5020 Brandin Ct.
Fremont, CA 94538
T: 510-498-4411
F: 510-683-2062
B: 510-226-9694

Exabyte Corporation   (*Distributor,*
Drives)
1685 38th St.
Boulder, CO 80301

T: 303-417-7792
F: 303-417-7890
 URL: http://www.exabyte.com/

Exar Corporation   (*ICs, Analog*)
48720 Kato Rd.
Fremont, CA 94539
T: 510-668-7000
F: 510-668-7017
 URL: http://www.exar.com/

Excalibur Communications
(*SW Utilities, Communication*)
T: 918-496-7881
B: 918-496-8113

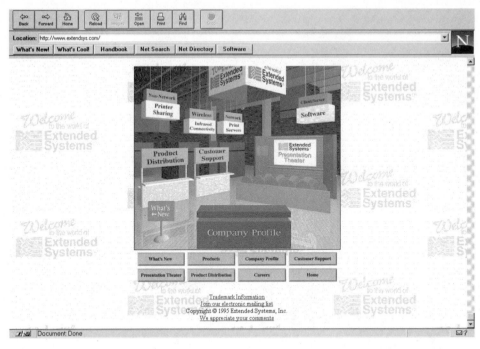

Figure 3-29 Extended Systems.

Exide (*Computers, Batteries*)
645 Penn St.
PO Box 14205
Reading, PA 19612-4205
T: 800-523-8954
URL:
http://www.exideworld.com/power/

Express Systems (*Computers, Systems*)
T: 800-321-4606
B: 206-728-8302
AOL: keyword EXPRESSSYSTEMS
CIS: go WINAPA
E-mail: info@express-systems.com

Extended Systems (*SW Utilities, Printer*)
5777 N. Meeker Ave.
Boise, ID 83713

T: 208-322-7800
F: 406-587-9170
FaxBack: 800-251-2612
E-mail: info@extendsys.com
URL: http://www.extendsys.com/

Extron (*Equipment, Test*)
1230 South Lewis St.
Anaheim, CA 92805
T: 714-491-1500
F: 714-491-1517
URL: http://www.extron.com/

Farallon Computing (*SW Utilities, Internet and Networking*)
2470 Mariner Square Loop
Alameda, CA 94501
T: 510-814-5000
AOL: keyword FARALLON

CIS: 75410,2702
E-mail: techsports@farallon.com
URL: http://www.farallon.com/

Fargo Electronics   (*Computers, Printers*)
7901 Flying Cloud Dr.
Eden Prairie, MN 55344
T: 612-941-9470
F: 612-941-7836
E-mail: support@fargo.com
URL: http://www.fargo.com/

Farnell Components   (*Distributor, Components*)
Canal Road, Leeds LS12 2TU UK
T: 0113 263 6311
F: 0113 263 3411
Tech Support: 0113 279 9123
DATALINE: 0113 231 0160
URL: http://www.farnell.co.uk/

FastComm Communications   (*Newtorking, Hardware*)
45472 Holiday Dr.
Sterling, VA 20166
T: 703-318-7750
F: 703-787-4625
E-mail: info@fastcomm.com
E-mail: support@fastcomm.com
FTP: ftp://ftp.fastcomm.com/
URL: http://www.fastcomm.com/

Ferran Scientific   (*Equipment, Instruments*)
11558 Sorrento Valley Rd., Suite 2
San Diego, CA 92121
T: 619-792-7549
F: 619-792-0065
URL: http://www.ferran.com/main.html

FG Commodity Electronics   (*Distributor, Components*)
3200 De Miniac
Ville St-Laurent, Quebec, H4S 1N5
Canada
T: 514-339-1425
F: 514-339-1429
E-mail: fred@fgcom.qc.ca
URL: http://www.fgcom.qc.ca/

Fibronics   (*Networking, Hardware*)
16 Esquire Rd.
North Billerica, MA 01862
T: 508-671-9440
F: 508-671-7262
URL: http://www.fibronics.co.il/

First Computer Systems   (*Distributor, Motherboards*)
6000 Live Oak Parkway, Ste 107
Norcross, GA 30093
T: 770-441-1911
F: 770-441-1856
E-mail: sales@fcsnet.com
URL: http://www.fcsnet.com/

Fisher   (*Peripherals, Sound*)
URL: http://www.audvidfisher.com/

FlashPak   (*Manufacturer, Custom*)
32 Bridge St
Elsternwick, Victoria, 3185 Australia
T: +61 3 528 2466
F: +61 3 528 5453
E-mail: cleve@werple.mira.net.au
URL: http://werple.mira.net.au/flashpak/

Fluke   (*Equipment, Test*)
PO Box 9090
Everett, WA 98206-9090
T: 206-347-6100

Figure 3-30   Fibronics.

F: 206-356-5116
URL: http://www.fluke.com/

Folio Corporation   (*SW Utilities,*
Business)
5072 N. 300 W.
Provo, UT 84604
T: 801-229-6700
FaxBack: 801-229-6390
B: 801-229-6668
URL: http://www.folio.com/

Forte   (*Networking, Software*)
1800 Harrison St.
Oakland, CA 94612
T: 510-869-3400
F: 510-869-3480
E-mail: support-agent@forteinc.com
E-mail: 3dpc.support@forteinc.com
URL: http://www.forte.com/

Fractal Design Corporation
(*SW Utilities, DTP*)
P.O. Box 2380
Aptos, CA 95001
T: 408-688-8800
F: 408-688-8836
URL: http://www.fractal.com/

Frame Technology Corporation
(*SW Utilities, DTP*)
333 West San Carlos St.
San Jose, CA 95110
T: 408-975-6466
F: 408-975-6611
B: 408-975-6729
FaxBack: 408-975-6731
CIS: go DTPVEND
FTP: ftp://ftp.frame.com/
URL: http://www.frame.com/

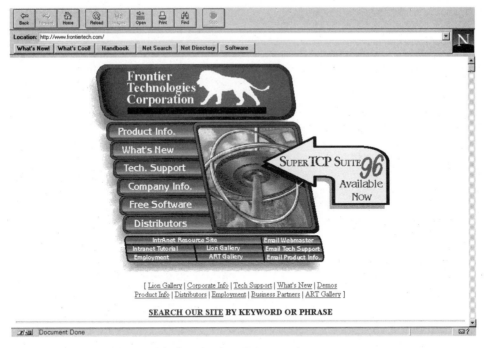

Figure 3-31 Frontier Technologies Corporation.

Frontier Technologies Corporation
(*SW Utilities, Internet*)
10201 N. Port Washington Rd.
Mequon, WI 53092
T: 414-241-4555
F: 414-241-7084
URL: http://www.frontiertech.com/

Frontline Design Automation
(*Simulation, Circuit CAD*)
2860 Zanker Rd., Suite 201
San Jose, CA 95134
T: 408-456-0222
F: 408-456-0265
URL: http://www.frontline.com/

FTP Software   (*SW Utilities,
Communication*)
100 Brickstone Sq.
Andover, MA 01810
T: 508-685-3300
E-mail: info@ftp.com
URL: http://www.ftp.com/

FTS Systems   (*Thermal Products, Heat
Sinks*)
3538 Main St.
PO Box 158
Stone Ridge, NY 12484-0158
T: 914-687-0071
F: 914-687-7481
URL: http://www1.mhv.net/~fts/
welcome.htm

Fujitsu America, Inc. (*ICs, Logic/
Computers, Systems*)
3055 Orchard Dr.
San Jose, CA 95134
T: 408-432-1300
F: 408-432-1318
B: 408-944-9899
URL: http://www.fujitsu.co.jp/
URL: http://www.fujitsu.com/

Funk Software, Inc. (*Networking,
Software*)
222 Third St.
Cambridge, MA 02142
T: 617-497-6339
F: 617-547-1031
FTP: ftp://www.funk.com/pub/
URL:
http://www.funk.com/wldetail.html

Future Electronics (*Distributor,
Components*)
237 Hymus Blvd.
Montreal (Pointe Claire), Quebec, H9R
5C7 Canada
T: 514-694-7710
URL: http://www.future.ca/

Galacticomm (*SW Utilities,
Communication*)
T: 954/583-5990
E-mail: sales@gcomm.com
URL: http://www.gcomm.com/

Galaxy Computers, Inc. (*Computers,
Systems*)
423 S. Lynnhaven Rd.
Virginia Beach, VA 23452
T: 800-809-1319
B: 804-486-5916
URL: http://www.galaxy-111.com/

Gammalink (*Computers, Modems*)
1314 Chesapeake Terrace
Sunnyvale, CA 94089
T: 800-FAX-4PCS
F: 408-744-1900
URL: http://www.gammalink.com/

Gateway 2000 (*Computers, Systems*)
PO Box 2000
610 Gateway Dr.
North Sioux City, SD 57049
T: 605-232-2000
F: 605-232-2023
B: 605-232-2109
FaxBack: 800-846-4526
AOL: keyword GATEWAY
CIS: go GATEWAY

General DataComm (*Networking,
Hardware*)
1579 Straits Turnpike
Middlebury, CT 06762-1299
Tel: 203-574-1118
F: 203-758-8507
URL: http://www.gdc.com/

General Electric (GE) (*Equipment,
Power Products*)
T: 800-626-2004
F: 518-869-5555
E-mail: geinfo@www.ge.com
URL: http://www.ge.com/   (Includes
links to all business units)

General Instrument Corporation
(*Equipment, Cable and Video*)
8770 W. Bryn Mawr Ave. Suite 1300
Chicago, IL 60631
T: 312-695-1000
URL: http://www.gi.com/

GigaTrend, Inc. (*Peripherals, Tape*)
2234 Rutherford Rd.
Carlsbad, CA 92008
T: 619-931-9122
F: 619-931-9959
B: 619-931-9469

Global Village Communication
(*Services, On-line*)
1144 East Arques Ave.
Sunnyvale, CA 94086
T: 408-523-1000
F: 408-523-2407
FaxBack: 408-523-2402
E-mail: sales@globalvillage.com
FTP: ftp://ftp.globalvillage.com/
URL: http://info.globalvillag.com/
welcome.html

GlobeTech International (*Equipment, Test*)
1684 South Research Loop
Tucson, Arizona, 85710
T: 602-298-6900
F: 602-298-1913
URL: http://www.gtinet.com/

Graychip (*ICs, DSP*)
2185-B Park Blvd.
Palo Alto, CA 94306
T: 415-323-2955
F: 415-323-0206
E-mail: sales@graychip.com
URL: http://www.best.com/~bevans/

Gupta (*Networking, Software*)
1060 Marsh Rd.
Menlo Park, CA 94025
T: 415-321-9500
URL: http://www.gupta.com/

HaL Computer Systems (*Computers, Workstations*)
1315 Dell Ave.
Campbell, CA 95008
T: 408-379-7000
F: 408-379-5022
URL: http://www.hal.com/

Halfin (*Distributor, Components*)
Quai Aux Pierres De Taille 37
B-1000 Bruxelles Belgium
T: +322/218-1011
F: +322/218-6998
URL:
http://www.wp.com/halfin/home.html

Hamilton-Hallmark (*Distributor, Components*)
T: 800-332-8638
URL: http://www.tsc.hh.avnet.com/
homepage.shtml

Hantro (*Simulation, Training*)
Hantro Products Oy
Teknologiantie 14, FIN-90570 Oulu,
Finland
T: +358-81-551 4382
Fax: +358-81-551 4490
E-Mail: info@hantro.pp.fi
URL:
http://www.pdlcomm.com/hantro/

Hard Drives Northwest (*Distributor, Drives*)
13407 NE 20th St.
Bellevue, WA 98005
T: 206-644-6474
F: 206-644-1963
E-mail: hdsales@hdnw.com
URL: http://www.hdnw.com/hduser/

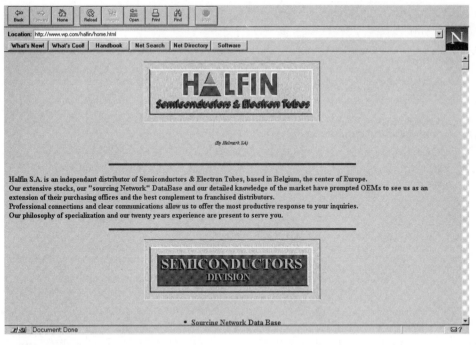

Figure 3-32   Halfin.

Harlequin   (*SW Utilities, DTP*)
E-mail: web@harlequin.com
URL: http://www.harlequin.com/
(Includes links to worldwide offices)

Harris Semiconductor   (*ICs, Mixed-Signal*)
T: 407-727-9207
E-mail: centapp@harris.com
URL: http://www.semi.harris.com/
(Includes links to worldwide offices)

Hartford Computer Group
(*Distributor, PC Assemblies*)
1610 Colonial Parkway
Inverness, IL 60067
T: 847-934-3380
F: 847-934-0157
URL: http://www.awa.com/hartford/

Hauppauge Computer Works, Inc.
(*Computers, Motherboards
Controllers, Video*)
91 Cabot Ct.
Hauppauge, NY 11788
T: 516-434-1600
F: 516-434-3198
B: 516-434-8454
URL: http://www.hauppauge.com/hcw/
index.htm

Hawk Electronics   (*Distributor, Components*)
T: 708-459-4030
F: 708-459-4091
E-mail: kip@hawkusa.com
URL: http://www.blol.com/hawk/
URL: http://www.hawkusa.com/

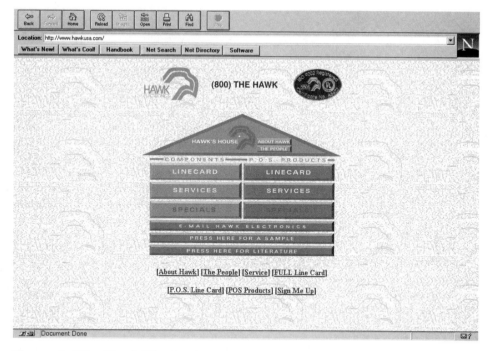

Figure 3-33 Hawk Electronics.

Hayes Microcomputer Products
(*Computers, Modems*)
5835 Peachtree Corners E.
Norcross, GA 30092-3405
T: 404-840-9200
F: 404-441-1213
B: 404-429-3734
CIS: go HAYES
Gopher: gopher://gopher.almac.co.
uk/11/business/comm/hayes/
URL: http://www.hayes.com/

HDS Network Systems, Inc. (*Computers, Workstations*)
400 Feheley Dr.
King of Prussia, PA 19406
T: 800-HDS-1551
E-mail: info@hds.com
URL: http://www.hds.com/

HDSS Computer (*Computers, Systems*)
T: 800-252-9777
URL: http://ccurrents.infinite.net/cc/
cci/gold/hdss/hdss2.html

Hercules Computer Technology
(*Controllers, Video*)
T: 510-623-6050
B: 510-623-7449
CIS: go HERCULES
E-mail: support@hercules.com
URL: http://www.dnai.com/~hercules/
URL: http://www.hercules.com/

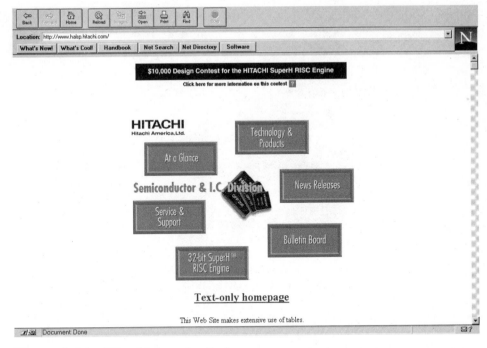

Figure 3-34   Hitachi America Ltd.

Heurikon Corporation   (*Computers, VME*)
8310 Excelsior Dr.
Madison, WI 53717
T: 608-831-5500
F: 608-831-4249
E-mail: info@heurikon.com
URL:
http://www.heurikon.com/index.html

Hewlett-Packard Company
(*Computers, Printers/Computers, Systems*)
16399 W. Bernardo Dr.
San Diego, CA 92127-1899
T: 619-592-4522
FaxBack: 208-344-4809
B: 408-720-3416
URL: http://www.hp.com/

Hilgraeve   (*SW Utilities, Communication*)
Genesis Centre
111 Conant Ave., Suite A
Monroe, MI 48161
T: 313-243-0576
F: 313-243-0645
B: 313-243-5915
CIS: go HILGRAEVE
E-mail: support@hilgraeve.com
URL: http://www.hilgraeve.com/

Hitachi America Ltd.   (*Computers, Drives/ICs, Logic*)
2000 Sierra Point Parkway
Brisbane, CA 94005
T: 800-285-1601
URL: http://www.halsp.hitachi.com/

URL: http://www.hii.hitachi.com/
URL: http://www.hitachi.co.jp/

Hi-Tech Component Distributors, Inc.
(*Distributor, PC Assemblies*)
59 S. La Patera Ln.
Goleta, CA 93117
T: 805-681-9961
F: 805-681-9971
URL: http://internet-café.com/hi-tech/

HiTech Equipment Corporation
(*Computers, Single Board*)
T: 619-566-1892
F: 619-530-1458
FTP: ftp://ftp.the.com/pub/

Hi-Tech Surplus   (*Distributor,
Components*)
605 East 44th St., Ste. 1B
Boise, ID 83714
T: 208-375-7516
F: 208-375-6571
URL: http://www.hitechsurplus.com/

HSC Software   (*SW Utilities, Imaging*)
T: 805-566-6200
AOL: keyword HSC
AOL: keyword KTP
CIS: go HSC
E-mail: kptsupport@aol.com

Hunter Technology Corporation
(*Manufacturer, Custom*)
2941 Corvin Dr.
Santa Clara, CA 95051
T: 408-245-5400
F: 408-736-1908
B: 408-245-5206
E-mail: Twin@ix.netcom.com
URL: http://www.nilva.com/hunter/
mainhunt.htm

Huntsville Microsystems, Inc. (HMI)
(*Equipment, Emulation*)
T: 205-881-6005
F: 205-882-6701
E-mail: sales@hmi.com
URL: http://www.hmi.com/

Hybrid Networks   (*Networking,
Cabling*)
10161 Bubb Rd.
Cupertino, CA 95014
T: 408-725-3250
F: 408-725-2439
E-mail: info@hybrid.com
URL: http://hybrid.com/

HyCalNet   (*Manufacturer, Sensors*)
T: 818-444-4000
F: 818-444-1314
E-mail: crc@hycalnet.com
URL: http://www.hycalnet.com/
sensors/

Hyperception   (*SW Utilities, DSP*)
9550 Skillman LB125
Dallas, TX 75243
T: 214-343-8525
F: 214-343-2457
B: 214-343-4108
E-mail: info@hyperception.com
URL: http://www.hyperception.com/

HyperDesk Corporation   (*SW Utilities,
Business*)
2000 West Park Dr.
Westboro, MA 01581
T: 508-685-3300
URL: http://www.hyperdesk.com/

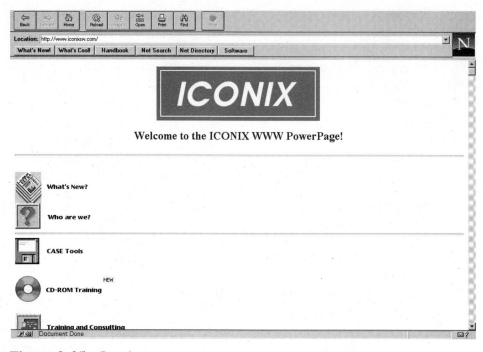

Figure 3-35 Iconix.

Hyundai Electronics America
(*Computers, Systems/Peripherals, Monitors*)
166 Baypointe Parkway
San Jose, CA 95134
T: 408-473-9200
URL: http://www.hea.com/

IBM (*Computers, Systems/ICs,*
Microprocessors)
T: 800-772-2227
FaxBack: 800-426-3395
B: 919-517-0001
AOL: keyword IBM
CIS: go IBMPS1, go IBMPS2, go
THINKPAD, go POWERPC,
go VALUEPOINT, go IBMSVR
E-mail: askibm@info.ibm.com
URL: http://www.csc.ibm.com/
URL: http://www.ibm.com/

URL: http://www.ibm.net/
URL: http://www.pc.ibm.com/

Iconix   (*SW Utilities, CASE*)
2800 28th St., Suite 320
Santa Monica, CA 90405
T: 310-458-0092
F: 310-396-3454
E-mail: marketing@iconixsw.com
URL: http://www.iconixsw.com/

Iconovex   (*SW Utilities, Internet*)
7900 Xerxes Ave. South
Bloomington, MN 55431
T: 612-896-5100
F: 612-896-5101
CIS: 74064,440
E-mail: FYI@iconovex.com
URL: http://www.iconovex.com/

ICS Learning Systems (*Services, Education*)
925 Oak St.
Scranton, PA 18515
T: 800-596-5505
URL: http://www.icslearn.com/

Imacon (*Distributor, Scanners*)
26 Hejrevej
DK-2400 Copenhagen NV, Denmark
Tel: +45 38 88 40 50
Fax: +45 38 88 40 52
URL: http://www.imacon.dk/

Imaging Magazine (*Publisher, Computers*)
12 West 21 St.
New York, NY 10010
URL:
http://www.imagingmagazine.com/

Imatek, Inc. (*Networking, Hardware*)
47490 Seabridge Dr.
Fremont, CA 94538
T: 510-683-8800
F: 510-683-8662
E-mail: router.support@empac.com
URL: http://www.imatek.com/

IMC Networks (*Networking, Hardware*)
16931 Millikan Ave.
Irvine, CA 92714
T: 714-724-1070
F: 714-724-1020
URL: http://www.imcnetworks.com/

IMP (*ICs, Logic*)
2830 North First St., MS 200
San Jose, CA 95134-2071
T: 408-432-9100
URL: http://www.impweb.com/

IMSI (*Computers, Multimedia*)
T: 415-257-3000 (Including FaxBack service)
B: 415-257-8468
E-mail: support@imsisoft.com

Industrial Programming, Inc.
(*SW Utilities, Operating Systems*)
100 Jericho Quadrangle
Jericho, NY 11753
T: 516-938-6600
F: 516-938-6609
E-mail: info@ipi.com
URL: http://www.ipi.com/

InfoChip Systems, Inc. (*Tools, Data Compression*)
2840 San Tomas Expressway
Santa Clara, CA 95051
T: 408-727-0514
B: 408-727-2496

Infolmaging Technologies, Inc.
(*SW Utilities, Imaging*)
3977 East Bayshore Rd.
Palo Alto, CA 94303
T: 415-960-0100
F: 415-960-0200
AOL: keyword 3DFAX
E-mail: info@infoimaging.com
URL: http://www.infoimaging.com/

Information Storage Devices
(*ICs, Logic*)
2045 Hamilton Ave.
San Jose, CA 95125
T: 408-369-2400
F: 408-369-2422
URL: http://www.isd.com/

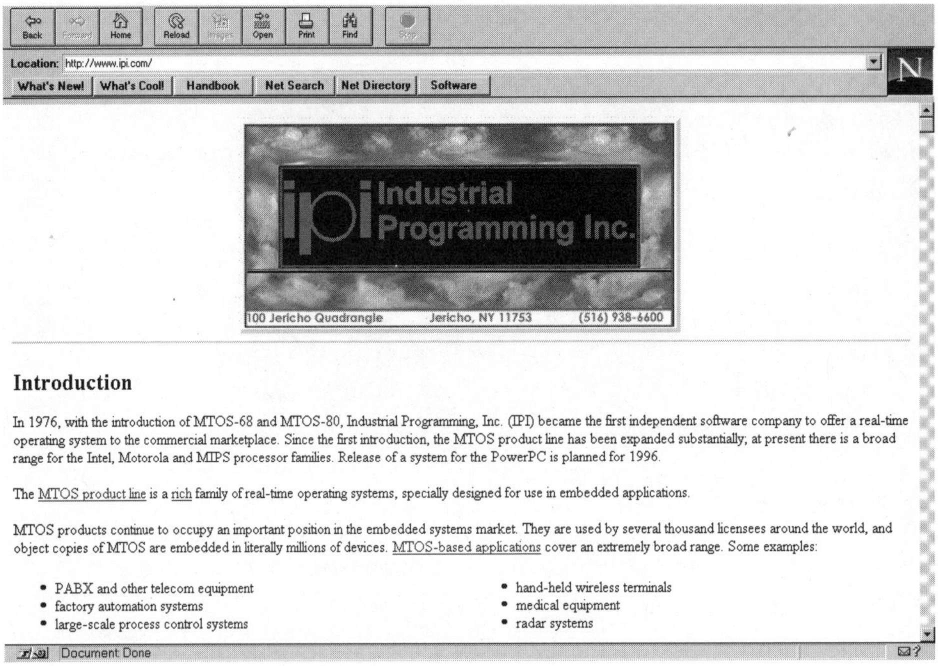

Figure 3-36 Industrial Programming Inc.

Infusion Systems, Ltd. (*Computers, Multimedia*)
1320 East Georgia St.
Vancouver, BC, V5L 2A8 Canada
T: 604-253-0747
F: 604-253-0747
E-mail: amulder@sfu.ca
URL: http://fas.sfu.ca/cs/people/ResearchStaff/amulder/personal/infusion/infusion.html

Ingram Micro (*Distributor, PC Assemblies*)
1600 E. St. Andrew Pl.
P.O. Box 25125
Santa Ana, CA 92799-5125
T: (714) 566-1000
URL: http://www.ingram.com/

Inset Systems (*SW Utilities, DTP*)
T: 203-775-5798
F: 203-740-1809
B: 203-740-0063
CIS: go INSET
E-mail: insetsupport@insetusa.com

Insight Direct (*Distributor, PC Assemblies*)
1912 West 4th St.
Tempe, AZ 85281
T: 800-848-9441
B: 602-902-5929
URL: http://www.insight.com/

Insight Electronics (*Distributor, Components*)
URL: http://www.ikn.com/
URL: http://memec.com/

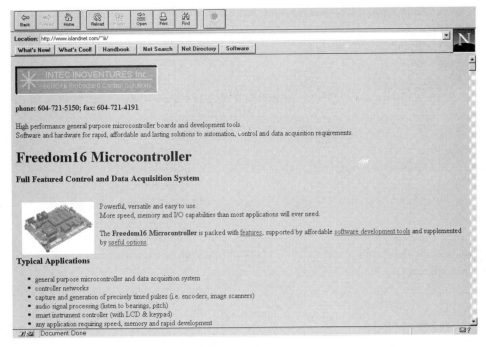

Figure 3-37   Intec Inoventures, Inc.

Insignia Solutions   (*SW Utilities, Internet*)
2200 Lawson Ln.
Santa Clara, CA 95054
T: 408-327-6000
F: 408-327-6105
URL: http://www.insignia.com/

Insulectro   (*Manufacturer, Custom*)
20362 Windrow Dr.
Lake Forest, CA 92630
T: 800-800-1003
URL: http://www.insulectro.com/

Intec Inoventures, Inc.   (*SW Utilities, Development*)
2751 Arbutus Rd.
Victoria, BC V8N 5X7 Canada
T: 604-721-5150
F: 604-721-4191

E-mail: iii@islandnet.com
URL: http://www.islandnet.com/~iii/

Integrated Circuit Design Concepts
(*Manufacturer, ICs*)
T: 714-633-0455
F: 714-838-7705
E-mail: icdcbob@netcom.com
FTP:
ftp://ftp.netcom.com/pub/ic/icdc/

Integrated Device Technology
(*ICs, Logic*)
2590 North First St., Bldg. A, Suite 100
San Jose, CA 95131
T: 408/943-9270
F: 408/943-9167
E-mail: info@idt.com
E-mail: techlit@idt.com
URL: http://www.idt.com/

Integrated Systems, Inc. (*Networking, Software*)
T: 408-980-1500
F: 408-980-0400
E-mail: psos_sales@isi.com
URL: http://www.isi.com/

Integrated Telecom Technology
(*ICs, Mixed-Signal*)
18310 Montgomery Village Ave.,
Suite 300
Gaithersburg, MD 20879
T: 301-990-9890
F: 301-990-9893
E-mail: products@igt.com
E-mail: techsupport@igt.com
URL: http://www.igt.com/

Intel Corporation
(*ICs, Microprocessors*)
200 Mission College Blvd.
Santa Clara, CA 95052
T: 408-765-7525
FaxBack: 916-356-3105
B: 916-356-3600
URL: http://www.intel.com/
URL: http://www.ssd.intel.com/
(Scaleable systems division)

Intelligent Instrumentation
(*Peripherals, Data Acquisition*)
6550 South Bay Colony Dr.
Tucson, AZ 86706
T: 520-573-3504
F: 520-573-0522
B: 520-294-7045
URL: http://www.instrument.com/

Interactive Image Technologies Ltd.
(*Simulation, Circuit CAD*)
908 Niagara Falls Blvd., #068
North Tonawanda, NY 14120-2060

T: 416-977-5550
F: 416-977-1818
B: 416-977-3540
CIS: 71333,3435
E-mail: ewb@interactiv.com
URL: http://www.interactiv.com/

Intercon (*SW Utilities, Internet*)
950 Herndon Parkway, Suite 400
Herndon, VA 22070
T: 703-709-5500
F: 703-709-5555
E-mail: info@intercon.com
E-mail: comment@intercon.com
URL: http://www.intercon.com/

Intergraph (*Simulation, Circuit CAD*)
T: 205-730-5441
URL: http://www.intergraph.com/
(Includes contacts to worldwide offices)
URL: http://www.veribest.com/sales/
data.html

Interlogic (*Computers, Industrial*)
85 Marcus Dr.
Melville, NY 11747
T: 516-420-8111
F: 516-420-8007
URL: http://www.infoview.com/

International Rectifier (*ICs, Analog*)
233 Kansas St.
El Segundo, CA 90245
T: 310-322-3331
F: 310-322-3332
URL: http://www.irf.com/~ir/

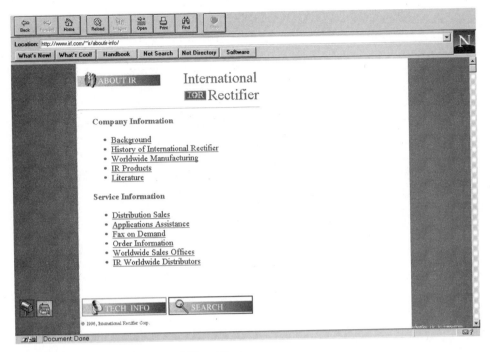

Figure 3-38 International Rectifier.

Interphase Corporation (*Controllers, Network*)
13800 Senlac
Dallas, TX 75234
T: 214-654-5000
F: 214-654-5500
E-mail: fastnet@iphase.com
URL: http://www.iphase.com/

Intusoft (*Simulation, Circuit CAD*)
Box 710
San Pedro, CA 90733-0710
T: 310-833-0710
URL: http://www.intusoft.com/

Iomega (*Computers, Drives*)
1821 West 4000 S.
Roy, UT 84067
T: 801-778-1000

F: 801-778-3461
B: 801-392-9819
AOL: keyword IOMEGA
E-mail: info@iomega.com
URL: http://www.iomega.com/

IQ Technologies, Inc. (*Controllers, Network*)
13625 NE 126th Pl.
Kirkland, WA 98034
T: 206-823-2273
F: 206-821-3961
B: 206-821-5486
E-mail: iq@blarg.com
URL: http://www.blarg.net/~iq/

ISDN Systems Corporation (ISC)
(*Networking, Hardware*)
8320 Old Courthouse Rd., Suite 200
Vienna, VA 22182
T: 703-883-0933
URL: http://www.infoanalytic.com/isc/
index.htm

ISG Technologies (*Equipment, Imaging*)
6509 Airport Rd.
Mississauga, ON, L4V 1S7 Canada
T: 905-672-2100
F: 905-672-2307
URL: http://www.isgtec.com/

I-Tech Corporation (*Equipment, Test*)
6975 Washington Ave. So., Ste. 220
Edina, MN 55343
T: 612-941-5905
B: 612-941-9610
E-mail: kolson@l-tech.com

ITU Technologies (*Equipment, Programming*)
3477 Westport Ct.
Cincinnati, OH 45248
T: 513-574-7523
F: 513-574-4245
E-Mail: sales@itutech.com
FTP: ftp://iglou.com/members/ITU/
URL: http://www.iglou.com/ITU/

Ivex Design International
(*Simulation, Circuit CAD*)
15232 NW Greenbrier Parkway
Beaverton, OR 97006-5746
T: 503-531-3555
F: 503-629-4907
B: 503-645-0576
E-mail: info@ivex.com
URL: http://www.ivex.com/

(Includes contacts for worldwide distributors)

Ivie Technologies (*Peripherals, Sound*)
1366 West Center St.
Orem, UT 84057
T: 801-224-1800
F: 801-224-7526
E-mail: lhdrisk@ivie.com
URL: http://sysint.com/systems/s4/
s4-1.htm

Jaco Electronics (*Distributor, Components*)
145 Oser Ave.
Hauppauge, NY 11788
T: 516-273-5500
F: 516-273-5506
URL: http://www.jacoelectronics.com/

Jameco Computer Products
(*Distributor, Components*)
1355 Shoreway Rd.
Belmont, CA 94002
T: 415-592-8097
F: 415-592-2503
B: 415-637-9025
E-mail: jameco@netcom.com

Janna Systems, Inc. (*SW Utilities, Business*)
CIS: go JANNA
URL: http://www.janna.com/

Jazz Multimedia (*Computers, Multimedia*)
T: 408-727-8900
B: 408-562-2010
AOL: keyword JAZZHELP
CIS: 75144,3536
E-mail: techsupp@jazzmm.com

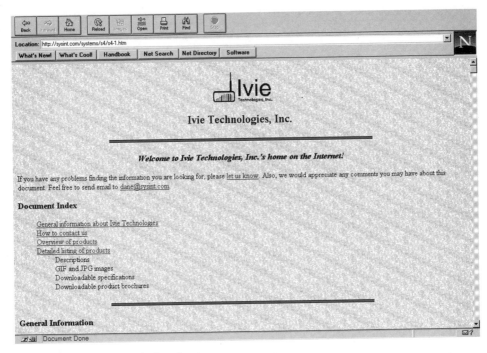

Figure 3-39 Ivie Technologies.

JDR Microdevices (*Distributor, Components*)
1850 South 10th St.
San Jose, CA 95112-4108
T: 800-538-5000
F: 800-538-5005
URL: http://www.jdr.com/

JECH Tech (*Equipment, Test*)
13962 Olde Post Rd.
Pickerington, OH 43147
T: 614-927-3495
F: 614-927-3494
URL:
http://www.infinet.com:80/~jectech/

Jem Computers (*Computers, Systems*)
35 Spinelli Place
Cambridge, MA 02138
T: 617-497-2500
F: 617-491-1006
E-mail: sales@jemcomputers.com
URL: http://www.tiac.net/biz/
jemsales/index.html

JMI Software Systems, Inc.
(*SW Utilities, Operating Systems*)
PO Box 481
904 Sheble Ln.
Spring House, PA 19477
T: 215-628-0840
F: 215-628-0353
E-mail: sales@jmi.com
URL: http://www.mcb.net/jmi/

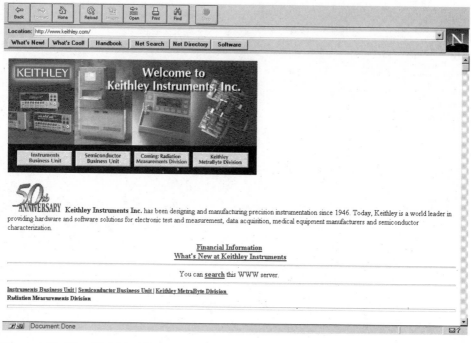

Figure 3-40 Keithley Instruments.

Kalpana (*Networking, Hardware*)
URL:
http://www.internex.com/kalpana/
URL: http://www.kalpana.com/

Kaman Instrumentation Corporation
(*Equipment, Instruments*)
1500 Garden of the Gods Rd.
Colorado Springs, CO 80907
T: 719-599-1132
F: 719-599-1823
E-mail: sensors@kamansensors.com
URL: http://www.rmii.com/kaman/

Keekor USA (*Manufacturer, Sensors*)
14806 N 74th St.
Scottsdale, AZ 85260
T: 602-443-0001
F: 602-443-0002
URL: http://www.keekor.com/

Keithley Instruments (*Equipment, Instruments*)
440 Myles Standish Blvd.
Taunton, MA 02780
T: 508-880-3000
F: 508-880-0179
FaxBack: 800-636-3377
E-mail: info@metrabyte.com
URL: http://www.keithley.com/

Keystone Learning Systems Corp.
(*Services, Education*)
2181 Larsen Parkway
Provo, UT 84606
T: 801-375-8680
F: 801-373-6872
URL: http://www.keylearnsys.com/

Key Tronic Corporation (*Peripherals, Input Devices*)
PO Box 14687
Spokane, WA 99214
T: 509-928-8000
F: 509-927-5248
B: 509-927-5288

Kingston Technology (*Distributor, Memory*)
17600 Newhope St.
Fountain Valley, CA 92708
T: 714-435-2600
F: 714-435-2699
B: 714-435-2636
URL: http://www.impediment.com/
URL: http://www.kingston.com/

K&K Associates (*Thermal Products, Software*)
10141 Nelson St.
Westminster, CO 80021
T: 303-702-1286
F: 303-772-5387
E-mail: takinfo@kkassoc.com
URL: http://www.csn.net:80/~takinfo/

Klever (*Networking, Hardware*)
1841 Zanker Rd.
San Jose, CA 95112
T: 408-467-0888
F: 408-467-0899
URL: http://www.klever.com/

KnowledgeBroker, Inc. (*SW Utilities, Business*)
T: 800-829-4KBI
URL: http://www.kbi.com/

Lamp Technology, Inc. (*Equipment, Industrial*)
1645 Sycamore Ave.
Bohemia, NY 11716-1729

T: 516-567-1800
F: 516-567-1806
URL: http://www.webscope.com/
lamptech/info.html

Lansdale Semiconductor
(*Manufacturer, ICs*)
2502 W. Huntington Dr.
Tempe, AZ 85282
T: 602-438-0123
F: 602-438-0138
E-Mail: lansdale@lansdale.com
URL: http://ssi.syspac.com/~lansdale/
index.html

LANSource (*Networking, Hardware*)
T: 416-535-3555
E-mail: sales@lansource.com

Larscom (*Networking, Hardware*)
4600 Patrick Henry Dr.
Santa Clara, CA 95054
T: 408-988-6600
F: 408-986-8690
E-mail: info@larscom.com
URL: http://www.larscom.com/

Laser Magnetic Storage (*Computers, Drives*)
4425 Arrowswest Dr.
Colorado Springs, CO 80907
T: 719-593-7900
F: 714-599-8713
B: 719-593-4081

Lattice Semiconductor (*ICs, Logic*)
1820 McCarthy Blvd.
Milpitas, CA 95035
T: 408-428-6400
F: 408-944-8450
URL: http://www.latticesemi.com/

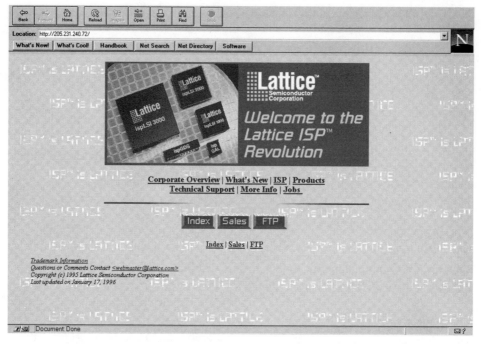

Figure 3-41 Lattice Semiconductor.

Lauterbach Datentechnik (*Equipment, Emulation*)
T: +49 8104-8943-0
F: +49-8104-8943-0
E-mail: sales@lauterbach.com
URL: http://www.lauterbach.com/

Lead Technologies (*SW Utilities, Imaging*)
T: 704-372-9681
F: 704-332-5868
B: 704-334-9045
CIS: go LEADTECH

Leapfrog Lab, Inc. (*Computers, Systems*)
47000 Warm Spring Blvd., #518
Fremont, CA 94539

T: 800-532-7550
F: 510-438-5788
E-mail: pc@leapfroglab.com
URL: http://www.leapfroglab.com/

LeCroy Corporation (*Equipment, Test*)
700 Chestnut Ridge Rd.
Chestnut Ridge, NY 10977-6499
T: 914-578-6013
F: 914-578-5984
URL: http://www.lecroy.com/

Level One Communications (*ICs, Communication*)
9750 Goethe Rd.
Sacramento, CA 95827
T: 916-855-5000
F: 916-854-1102
URL: http://www.level1.com/

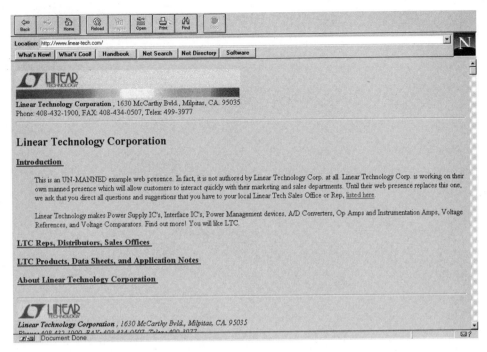

Figure 3-42 Linear Technology.

Lexmark International (*Computers, Printers*)
740 New Circle Rd.
Lexington, KY 40511
T: 606-232-2000
F: 606-232-3557
B: 606-232-5238
CIS: go LEXMARK
URL: http://www.lexmark.com/

Library Technologies, Inc. (LTI)
(*Simulation, IC Models*)
19959 Lanark Ln.
Saratoga, CA 95070
T: 408-741-1214
F: 408-867-2753
URL: http://www.libtech.com/

Likom (*Manufacturer, Custom*)
1751 McCarthy Blvd.
Milpitas, CA 95035
T: 408-954-8070
F: 408-954-8074
URL: http://www.likom.com.my/

Linear Technology (*ICs, Analog*)
1630 McCarthy Blvd.
Milpitas, CA 95035
T: 408-432-1900
F: 408-434-0507
URL: http://www.linear-tech.com/

Lots Technology, Inc. (*Peripherals, Tape*)
1556 Halford Ave. #113
Santa Clara, CA 95051
T: 408-747-1111
F: 408-747-0245
URL: http://www.win.net/~lasertape/welcome.html

Lotus Development Corporation
(*SW Utilities, Databases*)
55 Cambridge Pkwy.
Cambridge, MA 02142
T: 617-577-8500
F: 617-693-3899
B: 617-693-7000 (to 14.4KB)
CIS: go LOTUS
URL: http://www.lotus.com/

LSI Logic Corporation (*ICs, Logic*)
1551 McCarthy Blvd.
Milpitas, CA 95035
T: 408-433-8000
F: 408-433-7715
URL: http://www.Isilogic.com/

Lundahl Instruments (*Manufacturer, Sensors*)
T: 801-753-7300
F: 801-753-7490
E-mail: lundahl@cache.net
URL: http://www.cache.net/lundahl/

Macmillan Computer Publishing
(*Publisher, Computer*)
URL: http://www.mcp.com/
(Includes links to all publishing divisions)

Macromedia (*SW Utilities, Business*)
600 Townsend St.
San Francisco, CA 94103
T: 415-252-2000
F: 415-626-0554

AOL: keyword MACROMEDIA
CIS: go MACROMEDIA
URL: http://www.macromedia.com/

Madge Networks (*Networking, Enterprise*)
2310 North First St.
San Jose, CA 95131-1011
T: 408-955-0700
F: 408-955-0970
URL: http://www.madge.com/

Magnavox (*Peripherals, Monitors*)
One Phillips Dr.
Knoxville, TN 37914-1810
T: 423-521-4316
URL: http://www.magec.com/external.html
URL: http://www.magnavox.com/

Magnetic Technology (*Manufacturer, Tape Heads*)
20710 Manhattan Place, Suite 112
Torrance, CA 90501
T: 310-787-8094
F: 310-787-8096
URL: http://argus-inc.com/MagTech/MagTech.html

Mag-Tek (*Equipment, Magnetic Stripe*)
20725 South Annalee Ave.
Carson, CA 90746
T: 800-788-6835
F: 310-631-3956
URL: http://www.magtek.com/

Major League Electronics (*Computers, Connectors*)
4101 Reas Ln.
New Albany, IN 47150
T: 812-944-7244
F: 812-944-7268
Email: mle@iglou.com

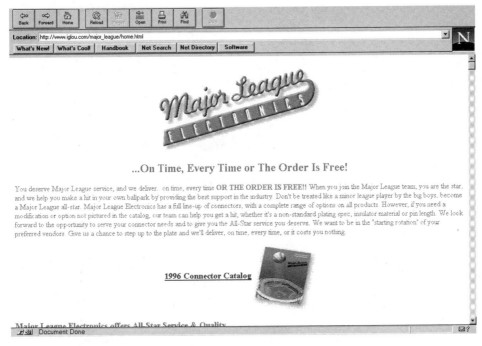

Figure 3-43   Major League Electronics.

URL:
http://www.iglou.com/major_league/

Marantz   (*Equipment, Sound*)
Glaslaan 2, Bldg. SFF-2
P.O. Box 80002
5600 JB Eindhoven, Netherlands
T: +31 40 2732241
F: +31 40 2735578
E-mail: marantz@am.umc.ce.philips.nl
URL: http://www.philips.com/marantz/

Marmah Magnetics, Inc.   (*Manufac-turer, Components*)
Marmah Magnetics Inc.
5450 Canotek Rd., Unit 74
Gloucester, Ontario, K1J-9G6 Canada
T: 613-747-0759
F: 613-747-0809

E-mail: marmag@magi.com.
URL: http://www.magi.com/
~marmag/marmah.html

Marsh Electronics   (*Distributor, Components*)
1563 South 101st St.
Milwaukee, WI 53214
T: 414-475-6000
F: 414-771-2847
URL: http://www.execpc.com/marsh/

Marshall Industries   (*Distributor, Components*)
9320 Telstar Ave.
El Monte, CA 91731-9937
E-mail: info@001.marshall.com
URL: http://www.marshall.com/

Marx International, Inc. (*Equipment, Encryption*)
20 Executive Park West, Suite #2027
Atlanta, GA 30329
T: 404-321-3020
F: 404-321-0760
CIS: 102543,2052
E-mail: marxint@ix.netcom.com
URL: http://electricpages.com/marx/

Mastersoft, Inc. (*SW Utilities, Business*)
8737 E. Via de Commercio
Scottsdale, AZ 85258
T: 602-277-0900
F: 602-948-8261
B: 602-596-5871

MathSoft (*SW Utilities, Scientific*)
101 Main St.
Cambridge, MA 02142-1521
T: 617-577-1017
F: 617-577-8829
CIS: go MATHSOFT
URL: http://www.mathsoft.com/

Mathsource (*SW Utilities, Scientific*)
Wolfram Research
100 Trade Center Dr.
Champaign, IL 61820-7237
T: 217-398-0700
F: 217-398-0747
E-mail: info@wolfram.com
URL:
http://www.wolfram.com/index1.html

MathWare (*SW Utilities, Scientific*)
P.O. Box 3025
Urbana, IL 61801
T: 800-255-2468
F: 217-384-7043
E-mail: mathware@xmission.com
URL: http://www.xmission.com/
~mathware/mathware.html

MathWorks (*SW Utilities, Scientific*)
24 Prime Park Way
Natick, MA 01760
T: 508-653-1415
F: 508-653-6284
E-mail: info@mathworks.com
URL: http://www.mathworks.com/

Matrox (*Controller, Video*)
1025 St. Regis Blvd.
Dorval, Quebec, H9P 2T4, Canada
T: 514-685-2630
F: 514-685-2853
E-mail: video.techsupport@matrox.com
(Technical Support)
E-mail: video.info@matrox.com (General
Information)
FTP Site:
ftp://ftp.matrox.com/pub/video/
URL: http://www.matrox.com/video

Matsushita Electric (*Computers, Drives*)
1006 Kadoma
Osaka, Japan
T: 06-908-1121
URL: http://www.mei.co.jp/index.html

Maynard Electronics, Inc. (*Peripherals, Tape*)
36 Skyline Dr.
Lake Mary, FL 32746
T: 407-263-3500
F: 407-262-4225
B: 407-263-3662

Maxim (*ICs, Analog*)
120 San Gabriel Dr.
Sunnyvale, CA 94086
URL: http://www.maxim-ic.com/

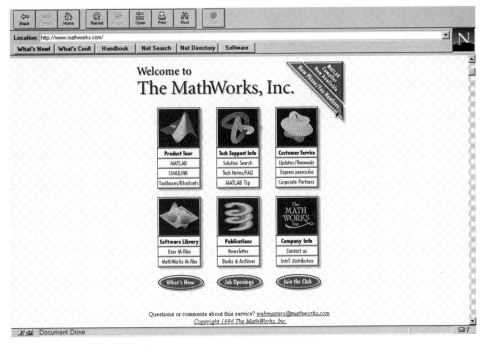

Figure 3-44 MathWorks.

Maximus Computers, Inc.
(*Computers, Systems*)
710 E. Cypress Ave., Unit A
Monrovia, CA 91016
T: 818-305-5925
F: 818-357-9140
B: 818-357-9790
URL: http://www.maximuspc.com/

Maxoptix (*Computers, Drives*)
3342 Gateway Blvd.
Fremont, CA 94538
T: 408-954-9700
F: 510-353-1845
B: 510-353-1448

Maxtor (*Computers, Drives*)
211 River Oaks Pkwy.
San Jose, CA 95134
T: 303-678-2700

F: 303-678-2585
B: 303-678-2020
FTP: ftp://ftp.maxtor.com/index.html
URL: http://www.maxtor.com/

Maxwell Laboratories, Inc.
(*Manufacturer, Custom*)
8888 Balboa Ave.
San Diego, CA 92123
T: 619-279-5100
F: 619-536-8035
URL: http://www.scubed.com/

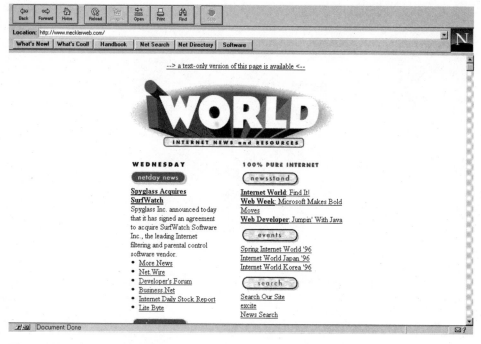

Figure 3-45   MecklerWeb.

McAfee Associates   (*Diagnostics, Virus*)
T: 408-988-3832
F: 408-970-9727
B: 408-988-4044
AOL: keyword MCAFEE
CIS: go MCAFEE
E-mail: support@mcafee.com
FTP: ftp.mcafee.com
URL: http://www.mcafee.com/

McGraw-Hill   (*Publishers, Computers*)
P.O. Box 545
Blacklick, OH 43004-0545
Gopher:
gopher://gopher.mcgraw.infor.com 5000
Telnet: mcgraw.infor.com
URL: http://mcgraw.infor.com:5000/

MECA Software   (*SW Utilities, Business*)
T: 203-255-7562
FaxBack: 203-256-9474
B: 203-259-2191

MecklerWeb   (*Services, On-line*)
20 Ketchum St.
Westport, CT 06880
T: 203-226-6967
F: 203-454-5840
E-mail: info@mecklermedia.com
URL: http://www.mecklerweb.com/

Mediatrix Peripherals   (*SW Utilities, Internet*)
4229 Garlock St.
Sherbrooke, Quebec, J1L 2C8 Canada
T: 819-829-8749
B: 819-829-5101

CIS: go MEDIATRIX
CIS: go MIDIV
E-mail: techsupp@mediatrix.com
URL: http://www.mediatrix.com/

Media Vision   (*Controller, Video*)
107 Bonaventura Dr.
San Jose, CA 95134
T: 541-882-1177
B: 510-770-0527
CIS: go MEDIAVISION
FTP: ftp://ftp.mediavis.com
URL: http://www.mediavis.com/

MegaHaus   (*Distributor, PC Assemblies*)
2201 Pine Dr.
Dickinson, TX 77539
T: 713-534-3919
F: 713-534-6580
E-mail: megahaus@phoenix.net
URL: http://www.megahaus.com/

Megahertz Corporation   (*Computers, Modems*)
605 North 5600 West
Salt Lake City, UT 84116
T: 801-320-7777
B: 801-320-8841
AOL: keyword MEGAHERTZ
CIS: go MEGAHERTZ
E-mail: techsupport@mhz.com
FTP: ftp://ftp.megahertz.com
URL: http://www.megahertz.com/
URL: http://www.xmission.com/~mhz/

MegaSoft   (*Services, On-line*)
819 Highway 33 East
Freehold, NJ 07728
T: 800-222-0490
F: 908-866-9376
E-mail: details@megasoft.com
URL: http://www.megasoft.com/

Mentor Graphics   (*SW Utilities, CAD*)
8005 SW Boeckman Rd.
Wilsonville, OR 97070-7777
T: 503-685-8000
F: 503-685-8001
URL: http://www.mentorg.com/

Merisel   (*Distributor, PC Assemblies*)
200 N. Continental Blvd.
El Segundo, CA 90245
T: 310-615-3080
F: 800-845-3744
B: 508-485-8507

Meta-Software   (*Simulation, Circuit CAD*)
1300 White Oaks Rd.
Campbell, CA 95008-6758
T: 408-369-5400
URL: http://www.metasw.com/

Methode Electronics   (*Manufacturer, Components*)
URL: http://www.methode.com/
(Includes links to all divisions)

Metz Software   (*SW Utilities, Communication*)
PO Box 6699
Bellevue, WA 98008-0699
T: 206-641-4525
F: 206-644-6026
B: 206-644-3663
AOL: keyword METZ
CIS: go METZ
URL: http://www.halcyon.com/metz/

MICOM   (*Networking, Hardware*)
4100 Los Angeles Ave.
Simi Valley, CA 93063-3397
T: 805-583-8600
F: 805-527-1031
URL: http://www.micom.com/

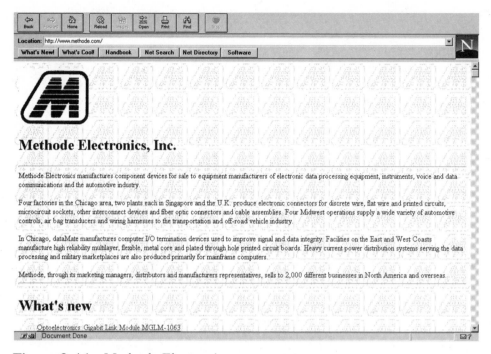

Figure 3-46 Methode Electronics.

Micro2000 (*Diagnostics, Hardware and Software*)
1100 E. Broadway, 3rd Floor
Glendale, CA 91205
T: 818-547-0125
F: 818-547-0397
URL: http://www.micro2000.com/

Microchip Technology (*ICs, RISC Processors*)
2355 West Chandler Blvd.
Chandler, AZ 85224-6199
T: 602-786-7200
URL:
http://www.mchip.com/microchip/
URL:
http://www.ultranet.com/biz/mchip/

Microcom (*Networking, Hardware and Software*)
500 River Ridge Dr.
Norwood, MA 02062
T: 617-551-1000
F: 617-551-1898
B: 617-255-1125
URL: http://www.microcom.com/

Micro Design International (MDI)
(*Controllers, SCSI*)
8985 University Blvd.
Winter Park, FL 32792
T: 407-677-8333
F: 407-677-8365
B: 407-677-4854

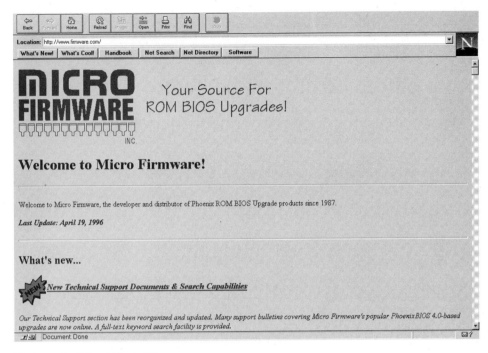

Figure 3-47 Micro Firmware.

Micro Digital, Inc. (*SW Utilities, Embedded*)
12842 Valley View St. #208
Garden Grove, CA 92645
T: 714-373-6862
F: 714-891-2363
B: 714-893-5118
E-mail: mdi@earthlink.net
URL: http://www.earthlink.net/~mdi/

MicroEdge (*SW Utilities, Development*)
PO Box 18038
Raleigh, NC 27619-8038
T: 919-831-0600
F: 919-831-0101
CIS: go SLICKEDIT
E-mail: sales@slickedit.com
URL: http://slickedit.com/

MicroExchange (*Computers, Refurbished*)
682 Passaic Ave.
Nutley, NJ 07110
T: 201-284-1200
E-mail: microexc@mmanet.com
URL:
http://www.mmanet.com/microexc/

MicroFirmware (*ICs, BIOS*)
330 W. Gray St., Suite 170
Norman, OK 73069-7111
T: 405-321-8333
F: 405-573-5535
B: 405-321-3553
E-mail: support@firmware.com
URL: http://www.firmware.com/

Micro Focus, Ltd. (*SW Utilities, Development*)
2465 East Bayshore Rd.
Palo Alto, CA 94303
T: 415-856-4161
F: 415-856-6134
URL:
http://www.mfltd.co.uk/win95.html

Micrografx (*SW Utilities, CAD*)
1303 E. Arapaho Rd.
Richardson, TX 75081
T: 214-234-1769
F: 214-994-6476
B: 214-644-4194
AOL: keyword MICROGRAFX
CIS: go MICROGRAFX
URL: http://www.micrografx.com/

Micro House International
(*SW Utilities, Reference*)
4900 Pearl East Circle, Ste. 101
Boulder, CO 80301
T: 303-443-3389
F: 303-443-3323
B: 303-443-9957
URL: http://www.microhouse.com/

Micro Logic (*SW Utilities, Business*)
PO Box 70
Hackensack, NJ 07602
T: 201-342-6518
F: 201-342-0370
E-mail: staff@miclog.com
URL: http://www.miclog.com/

MicroMath Scientific Software
(*SW Utilities, Scientific*)
2469 E. Ft. Union Blvd., Ste. 200
PO Box 71550
Salt Lake City, UT 84171-0550
T: 801-943-0290

F: 801-943-0299
B: 801-943-0397
URL: http://www.MicroMath.com/

Micron Electronics, Inc. (*Computers, Systems*)
900 E. Karcher Rd.
Nampa, ID 83687
T: 208-463-3434
F: 208-463-3424
URL: http://www.mei.micron.com/

Micronics Computers (*Computers, Motherboards*)
232 E. Warren Ave.
Fremont, CA 94539
T: 510-651-2300
F: 510-651-5612
B: 510-651-6837

Micron Technology (*ICs, Logic*)
8000 S. Federal Way
Boise, ID 83707
T: 208-368-3900
F: 208-368-3809
B: 208-368-4530
URL: http://www.micron.com/
URL:
http://www.micron.com/mti/index.htm

Micropolis Corporation (*Computers, Drives*)
100 Century Center Ct., Suite 410
San Jose, CA 95112-4512
T: 408-441-0333
F: 408-441-0334
B: 818-709-3310
URL: http://www.microp.com/

Microsemi (*Manufacturer, Components*)
2830 S. Fairview St.
Santa Ana, CA 92704

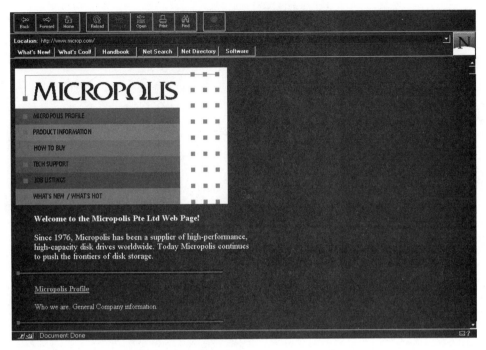

Figure 3-48 Micropolis Corporation.

T: 714-979-8220
F: 714-557-5989
URL: http://www.microsemi.com/
(Includes links to various divisions)

MicroSim Corporation (*Simulation, Circuit CAD*)
20 Fairbanks
Irvine, CA 92718
T: 714-770-3022
F: 714-455-0554
B: 714-830-1550
URL: http://www.microsim.com/

Microsoft Corporation (*SW Utilities, Operating Systems*)
CIS: go MICROSOFT
URL: http://www.microsoft.com/

URL: http://www.microsoft.com/
windows/ie/ie.htm
(the Internet browser for Windows 95)
URL: http://www.microsoft.com/kb/
softlib/windows/windows95/
windows95.htm
(the Microsoft Windows 95 software
library)
URL: http://www.msn.net/

Micro Solutions (*Distributor, Drives*)
132 W. Lincoln Hwy
DeKalb, IL 60115
T: 815-756-3411
F: 815-756-2928
B: 815-756-9100
URL: http://www.micro-solutions.com/

MicroSystems Development
(*Diagnostics, Software*)
4100 Moorpark Ave., Suite 104
San Jose, CA 95117
T: 408-269-4100
F: 408-296-5877
B: 408-296-4200

Micro Technology Services
(*Manufacturer, Custom*)
1819 Firman Dr. #137
Richardson, TX 75081
T: 214-231-6874
F: 214-669-1599
URL: http://www.mitsi.com/

Microtek International (*SW Utilities, Development*)
T: 503-645-7333
F: 503-629-8460
E-mail: info@microtekintl.com
URL: http://www.microtekintl.com/

Microtronix Datacomm, Ltd.
(*Communications, ISDN*)
200 Aberdeen Dr.
London, Ontario N5V 4N2 Canada
T: 519-659-9500
F: 519-659-8500
URL: http;//www.microtronix.com/

Micro-Vision (*Services, Repair*)
T: 813-527-3388
B: 813-526-3388
E-mail: microv@intnet.net
URL: http://www1.trib.com/ADS/
MICROV/

Micro Warehouse (*Distributor, Software*)
T: 908-370-4779
F: 908-905-1135

E-mail: service@warehouse.com
URL: http://www.warehouse.com/

Microwave Filter Co., Inc.
(*Manufacturer, Components*)
6743 Kinne St.
E. Syracuse, NY 13057
T: 315-437-3953
F: 315-463-1467
Email: mfc@ras.com
URL: http://rway.com/mwfilter/

Microwave Power, Inc.
(*Manufacturer, Components*)
3350 Scott Blvd, Bldg. 25
Santa Clara, CA 95054
T: 408-727-6666
F: 408-727-2246
E-mail: info@microwavepwr.com
URL: http://www.microwavepwr.com/

MicroXperts (*Distributor, PC Assemblies*)
T: 800-214-4029
URL: http://www.microx.com/

Midwest Micro (*Computers, Systems*)
6910 US Rt. 36 East
Fletcher, OH 45326
T: 800-262-6622
F: 800-445-2025
URL: http://www.mwmicro.com/

Milbert Amplifiers, Inc. (*Peripherals, Sound*)
Box 1027
Germantown, MD 20874
T: 301-963-9355
F: 301-840-0511
E-mail: mmilbert@access.digex.net
URL: http://www.access.digex.net:80/
~mmilbert/

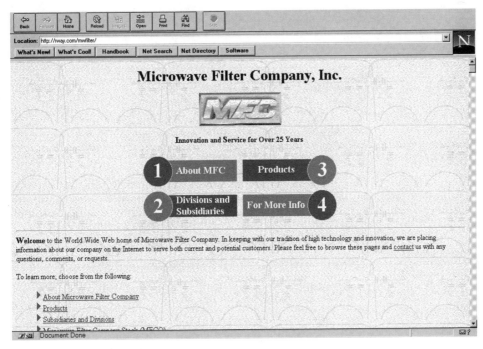

Figure 3-49 Microwave Filter Co., Inc.

Milplex Circuits, Inc. (*Manufacturer, PCB*)
1301 W. Ardmore Ave.
Itasca, IL 60143-1104
T: 708-250-1580
F: 708-250-1590
E-mail: sales@milplex.com
URL: http://milplex.com/

Miltec Electronics, Inc.
(*Manufacturer, PCB*)
7 Jones St.
Norcross, GA 30071-2503
T: 404-662-5922
F: 404-448-5016
B: 404-242-0266
E-mail: info@miltec.com
URL: http://www.mindspring.com/
miltec/

MiniMicro (*Distributor, PC Assemblies*)
2050 Corporate Ct.
San Jose, CA 95131
T: 408-456-9500
F: 408-434-9242
B: 408-434-9319
E-mail: postmaster@mm.sco.com

MIPS Technologies (*ICs, RISC Processors*)
2011 N. Shoreline Blvd.
P.O. Box 7311
Mountain View, CA 94039
T: 800-998-6477
F: 415-688-4321
URL: http://www.mips.com/

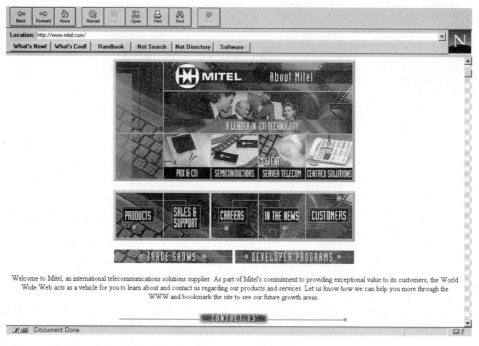

Figure 3-50   Mitel Semiconductor.

miro Computer Products   (*Equipment, Video*)
T: 800-549-6476
B: 415-855-9944

Mission Electronics   (*Distributor, Memory*)
T: 800-472-6233
URL: http://www.dram.com/

Mitel Semiconductor   (*ICs, Communication*)
P.O. Box 13089
360 Legget Dr.
Kanata, Ontario K2K 1X3 Canada
T: 613-592-2122
F: 613-592-6909
URL: http://www.mitel.com/
URL: http://www.semicon.mitel.com/

Mitronics   (*Distributor, Components*)
T: 201-812-2665
F: 201-812-2692
E-mail: mitron@mitronics.com
URL: http://w3.gti.net:80/2/

Mitsubishi Electronics America
(*Computers, Drives*)
T: 800-344-6352
FaxBack: 714-236-6453
B: 714-236-6286
URL: http://www.mitsubishi.co.jp/
(Includes links to all business units)

MKS (Mortice Kern Systems)
(*SW Utilities, Development*)
185 Columbia St. West
Waterloo, Ontario, N2L 5Z5, Canada
T: 519-884-2270
F: 519-884-8861

B: 519-884-2861
E-mail: support@mks.com
URL: http://www.mks.com/

MobileWare (*SW Utilities,*
Communication)
2425 North Central Expressway, Suite 120
Richardson, TX 75080-2748
T: 214-952-1200
F: 214-690-6185
E-mail: info@mobileware.com
URL: http://www.mobileware.com/

Model Technology (*Simulation,*
Circuit CAD)
8905 SW Watkins Ave. Suite 150
Beaverton, OR 97008
T: 503-641-1340
F: 503-526-8742
URL: http://www.model.com/

**Molecular OptoElectronics Corpora-
tion** (*Manufacturer, Components*)
877 25th St.
Watervliet, NY 12189
T: 518-270-8208
E-mail: sales@moec.com
URL:
http://www.automatrix.com/moec/

Molex, Inc. (*Computers, Connectors*)
2222 Wellington Ct.
Lisle, IL 60532
T: 708-969-4550
F: 708-968-8356
E-mail: amerinfo@molex.com
URL: http://www.molex.com/

Moon Valley Software (*SW Utilities,*
Internet)
141 Suburban Rd., Suite A1
San Luis Obispo, CA 93401

T: 805-781-3890
F: 805-781-3898
E-mail: info@moonvalley.com
URL: http://www.moonvalley.com/

Morning Star Technologies
(*SW Utilities, Internet*)
3518 Riverside Dr., Suite 101
Columbus OH 43221-1754
T: 614-451-1883
F: 614-459-5054
E-mail: Support@MorningStar.Com
FTP: ftp://ftp.MorningStar.Com/pub/
URL: http://www.morningstar.com/

Motherboard Discount Center
(*Distributor, Motherboards*)
1035 N. McQueen, Ste 123
Gilbert, AZ 85233
T: 602-813-6547
F: 602-813-8002
E-mail: mdc@primenet.com
URL: http://www.homepage.com/
mrupgrade/

Motion Pixels (*SW Utilities, Video*
Compression)
7320 E. Butherus Dr., #100
Scottsdale, AZ 85260
T: 602-951-3288
F: 602-951-2683
E-mail: questions@siriuspub.com
URL: http://www.motionpixels.com/

Motorola (*ICs, Microprocessors/*
Equipment, Communication)
URL: http://www.mot.com/
(Includes links to all business units)
URL: http://motserv.indirect.com/

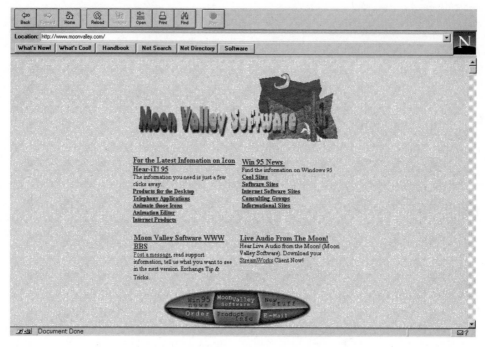

Figure 3-51   Moon Valley Software.

Mountain Network Solutions, Inc.
(*Peripherals, Tape*)
360 El Pueblo Rd.
Scotts Valley, CA 95066
T: 800-458-0300
F: 408-438-7623
B: 408-438-2665

Mouser Electronics   (*Distributor, Components*)
12 Emery Ave.
Randolph, NJ 07869
T: 201-328-3322
F: 201-328-7120
E-mail: sales@mouser.com
URL: http://www.mouser.com/

MPC Technologies, Inc.   (*Computers, Systems*)
2915 Daimler St.
Santa Ana, CA 92705
T: 714-724-9000
F: 714-724-9648
B: 714-724-1519
E-mail: help@mpctech.com

MPR Teltech   (*Networking, Hardware*)
8999 Nelson Way
Burnaby, BC, V5A 4B5, Canada
T: 604.294.1471
F: 604.293.5787
URL: http://www.mpr.ca/

Mr. BIOS   (*ICs, BIOS*)
Microid Research, Inc.
2336-D Walsh Ave.
Santa Clara, CA 95051, USA

Figure 3-52   Music Semiconductors.

email: mrbios@mrbios.com
URL: http://www.mrbios.com/

Multiple Brand Superstore (MBS)
(*Distributor, PC Assemblies*)
10976 S. Richardson Rd.
Ashland, Virginia 23005
T: 800-944-3808
F: 804-550-0680
URL: http://www.mbss.com/

Music Semiconductors   (*ICs, Logic*)
1150 Academy Park Loop, Suite 202
Colorado Springs, CO 80910
T: 719-570-1550
F: 719-570-1555
URL: http://www.music.com/

Mustang Software   (*SW Utilities,
Communication*)
PO Box 2264
Bakersfield, CA 93303
T: 805-873-2500
F: 805-873-2599
B: 805-873-2400
FTP: ftp://ftp.mustang.com/
URL: http://www.mustang.com/

MX.COM, Inc.   (*ICs, Mixed Signal*)
T: 800-777-7334
URL: http://www.compdist.com/
mxcom/index.html

Mylex   (*Controllers, SCSI*)
34551 Ardenwood Blvd.
Fremont, CA 94555
E-mail: support@mylex.com
URL: http://www.mylex.com/

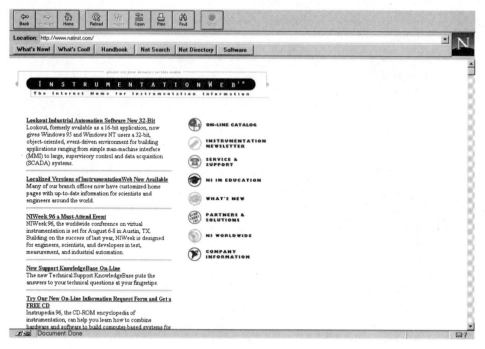

Figure 3-53 National Instruments.

Nanao USA (*Peripherals, Monitors*)
23535 Telo Ave.
Torrance, CA 90505
T: 310-325-5202
FaxBack: 800-416-3539
URL: http://www.traveller.com/nanao/

Nanothinc (*Services, Information*)
URL: http://www.nanothinc.com/

Napersoft, Inc. (*SW Utilities, Business*)
T: 708-420-1515
E-mail: info@napersoft.com
URL: http://www.napersoft.com/

National Computer Tectonics
(*Publisher, Internet*)
Markland Communities, Inc.
451 Blackburn Ln.
Lenoir City, TN 37771

URL: http://da.awa.com/nct/

National Instruments (*Equipment, Instruments*)
6504 Bridge Point Parkway
Austin, TX 78730
T: 512-794-0100
F: 512-794-8411
B: 512-794-5422
E-mail: info@natinst.com
FTP: ftp://ftp.natinst.com
URL: http://www.natinst.com/

National MicroComputers (*Computers, Systems*)
B: 801-261-1102

National Parts Depot (*Distributor, Components*)
31 Elkay Dr.
Chester, NY 10918
T: 914-469-4800
F: 914-469-4855
URL: http://www.megasoft.com/npd/

National Semiconductor (*ICs, Analog*)
2500 Semiconductor Dr.
Santa Clara, CA 95052-8090
T: 408-721-4299
F: 408-721-7582
B: 408-245-0671
URL: http://www.natmeni.com/
URL: http://www.nsc.com/

NCR Microelectronics (*ICs, Logic*)
1635 Aeroplaza
Colorado Springs, CO 80916
T: 719-596-5795
F: 719-573-3286
B: 719-574-0424

NEC Technologies, Inc. (*Peripherals, Monitors*)
1414 Massachusetts Ave.
Boxborough, MA 01719
T: 508-264-8000
F: 508-264-8245
B: 708-860-2602
FTP: ftp.nec.com
FTP: ftp.nectech.com
FTP: ftp.rascal.ics.utexas.edu/~ftp/mac/support-of-products/nec/
URL: http://www.nec.com/

Needham's Electronics (*Equipment, Programming*)
4630 Beloit Dr., Ste. 20
Sacramento, CA 95838
T: 916-924-8037
F: 916-924-8065

B: 916-924-8094
FTP: ftp://ftp.crl.com/users/ro/needhams/
URL: http://www.crl.com/~needhams/
URL: http://www.quiknet.com/~needhams/

Nesbit Software (*SW Utilities, Internet*)
4367 S.W. Spratt Way, Ste. 304
Beaverton, OR 97007
E-mail: kenn@nesbitt.com
E-mail: techsupt@nesbitt.com
URL: http://www.nesbitt.com/

NetCom (*Services, On-line*)
3031 Tisch Way
San Jose, CA 95128
T: 408-983-5950
F: 408-241-9145
E-mail: info@netcom.com
URL: http://www.netcom.com/net-com/homepage.html

NetEdge (*Networking, Hardware*)
PO Box 14993
Research Triangle Park, NC 27709
T: 800-NET-EDGE
F: 919-991-9160
URL: http://www.netedge.com/

NetManage, Inc. (*SW Utilities, Internet*)
10725 N. De Anza Blvd.
Cupertino, CA 95014
T: 408-973-7171
B: 408-257-3794
E-mail: support@netmanage.com
URL: http://www.netmanage.com/

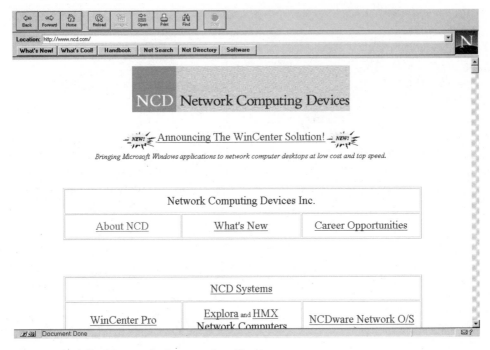

Figure 3-54   Network Computing Devices.

Netrix Corporation   (*Networking, Hardware*)
T: 703-742-6000
F: 703-742-4049
B: 703-793-1233
URL: http://www.netrix.com/

Network Computing Devices (NCD)
(*Networking, Hardware and Software*)
350 North Bernardo Ave.
Mountain View, CA 94043-5207
T: 415-694-0650
F: 415-961-7711
E-mail: info@ncd.com
URL: http://www.ncd.com/

Netscape Communications Corporation   (*SW Utilities, Internet*)
501 E. Middlefield Rd.
Mountain View, CA 94043
T: 415-528-3777
F: 415-528-4124
URL: http://www.netscape.com/

Newbridge   (*Networking, Hardware*)
P.O. Box 13600, 600 March Rd.
Kanata, Ontario, K2K 2E6, Canada
T: 613-591-3600
F: 613-591-3680
URL: http://www.newbridge.com/
URL: http://www.vivid.newbridge.com/

NexGen, Inc.   (*ICs, Microprocessors*)
1623 Buckeye Dr.
Milpitas, CA 95035

T: 408-325-8283
B: 408-955-1839
FTP: ftp://ftp.nexgen.com
URL: http://www.nexgen.com/

NeXT Computer (*SW Utilities, Internet*)
900 Chesapeake Dr.
Redwood City, CA 94063
T: 415-366-0900
F: 415-780-3929
URL: http://www.next.com

NMB Technologies (*Peripherals, Input Devices*)
9730 Independence Ave.
Chatsworth, CA 91311
T: 818-341-3355
F: 818-341-8207
URL: http://www.nmbtech.com/

Nokia Display Products (*Peripherals, Monitors*)
1505 Bridgeway Blvd, Suite 128
Sausalito, CA 94965
T: 415-331-4244
F: 415-331-0424
URL: http://www.nokia.com/
URL: http://www.sjmercury.com/advert/nokia/

Northern Telecom (*Communications, Hardware*)
2010 Corporate Ridge
McLean, VA 22102
T: 703-712-8000
URL: http://www.nortel.com/

Northgate Computer Systems, Inc.
(*Computers, Systems*)
10025 Valley View Rd., #110
Eden Prairie, MN 55344

T: 800-548-1993
F: 612-947-4604
B: 612-947-4640

Novell, Inc. (*Networking, Software*)
122 E. 1700 South
Provo, UT 84601
T: 801-379-5588
F: 801-429-5157
B: 801-429-3030
FTP: ftp.novell.com
URL: http://www.novell.com/
URL: http://www.netware.com/

NTT (Nippon Telegraph & Telephone)
(*Communications, Hardware*)
URL: http://www.ntt.jp/
(Includes links to all business units)

Nu Horizons Electronics (*Distributor, Components*)
6000 Nu Horizons Blvd.
Amityville, NY 11701
T: 516-226-6000
F: 516-226-6140
E-mail: sales@nuhorizons.com
URL: http://www.nuhorizons.com/

Number Nine Visual Technology
(*Controllers, Video*)
18 Hartwell Ave.
Lexington, MA 02173-3103
T: 617-674-0009
F: 617-674-2919
B: 617-862-7502
CIS: go NINE
URL: http://www.nine.com/

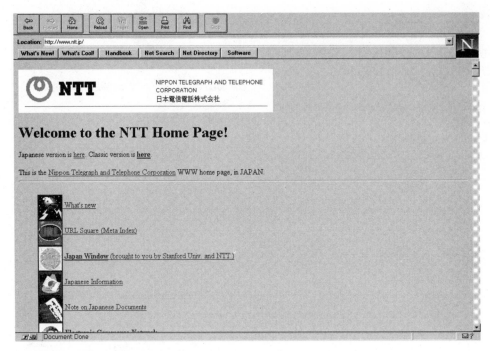

Figure 3-55   NTT.

Nu-Mega Technologies, Inc.
(*SW Utilities, Development*)
P.O. Box 7780
Nashua, NH 03060-7780
T: 603-889-2386
F: 603-889-1135
E-mail: info@numega.com
URL: http://www.numega.com/

Numera Software   (*SW Utilities, CAD*)
1501 Forth Ave., Ste. 2880
Seattle, WA 98101
T: 206-622-2233
F: 206-622-5382
CIS: 74774,532
URL: http://www.numera.com/

Nu-Tronic Circuit Company
(*Manufacturer, PCB*)
250 East 17th St.
Paterson, NJ 07524
T: 201-279-3838
F: 201-684-8940
E-mail: sales@nutron.com
URL: http://www.nutron.com/

NVidia Corporation   (*ICs, Logic*)
1226 Tiros Way
Sunnyvale, CA 94086
T: 408-720-6100
F: 408-720-6111
E-mail: info@nvidia.com
URL: http://www.nvidia.com/

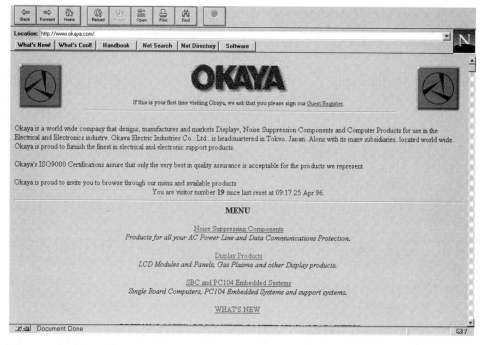

Figure 3-56   Okaya Electric America, Inc.

ObjecTime Limited   (*SW Utilities, Development*)
340 March Rd.
Kanata, Ontario, K2K 2E4, Canada
T: 613-591-3535
F: 613-591-3784
E-mail: sales@objectime.on.ca
URL: http://www.objectime.on.ca/

Occo Enterprises, Inc.   (*Computers, Systems*)
470 Cloverleaf Dr., Unit B
Baldwin Park, CA 91706
T: 818-333-2222
F: 818-333-1111
URL: http://www.ibv.com/occo/

Odyssey Computing   (*Computers, Systems*)
T: 619-675-3775
CIS: go ODYSSEY
E-mail: support@odysseyinc.com

Okaya Electric America, Inc.
(*Peripherals, LCDs*)
503 Wall St.
Valparaiso, IN 46383
T: 219-477-4488
F: 219-477-4856
URL: http://www.okaya.com/

Oki America, Inc. (*Manufacturer, Components*)
URL:
http://www.oki.co.jp/OKI/English/
URL: http://www.oki.com/
(Includes links to all business units)
URL: http://www.okisemi.com/

Okidata (*Peripherals, Printers*)
532 Fellowship Rd.
Mt. Laurel, NJ 08054
T: 609-235-2600
F: 609-424-7423
B: 800-283-5474
URL: http://www.okidata.com/

Olivetti (*Peripherals, Printers*)
765 U.S. Highway 202
P.O. Box 6945
Bridgewater NJ 08807-0945
T: 908-526-8200
F: 908-218-5677
URL: http://www.isc-br.com/

Omni Data Communications
(*Distributor, PC Assemblies*)
906 N. Main, Suite 3
Wichita, KS 67203
T: 316-264-5068
F: 316-264-7031
Tech Support: 316-264-5589
E-mail: odcom@aol.com
URL: http://www.omnidata.com/

Ontrack Computer Systems, Inc.
(*Services, Data Recovery Diagnostics, Software*)
6321 Bury Dr., Suites 15-19
Eden Prairie, MN 55346
T: 612-937-2121
F: 612-937-5750

B: 612-937-8567
URL: http://www.ontrack.com/

Open Market (*SW Utilities, Internet*)
245 First St.
Cambridge, MA 02142
T: 617-621-9500
F: 617-621-1703
E-mail: sales@openmarket.com
URL: http://www.openmarket.com/

OPTI, Inc. (*ICs, Logic*)
2525 Walsh Ave.
Santa Clara, CA 95051
T: 408-980-8178
F: 408-980-8860
B: 408-980-9774
URL: http://www.opti.com/

Optotek (*Simulation, Circuit CAD*)
62 Steacie Dr.
Kanata, Ontario, K2K 2A9 Canada
T: 613-591-0336
F: 613-591-0584
E-mail: optotek@optotek.com
URL: http://www.optotek.com/

Oracle (*SW Utilities, Databases*)
500 Oracle Parkway
Redwood Shores, CA 94065
T: 415-506-7000
F: 415-506-7200
URL: http://www.oracle.com/

OrCad (*Simulation, Circuit CAD*)
9300 S.W. Nimbus Ave.
Beaverton, OR 97008
T: 503-671-9500
F: 503-671-9501
E-mail: info@orcad.com
URL: http://www.orcad.com/

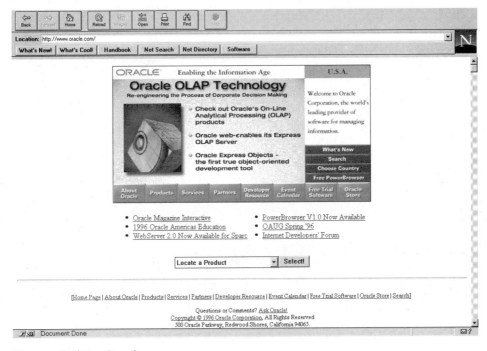

Figure 3-57 Oracle.

Orchid Technology (*Controllers, Video*)
45365 Northport Loop West
Fremont, CA 94538
T: 510-683-0300
F: 510-651-5612
B: 510-683-0329
URL: http://www.orchid.com/

Orion Instruments (*SW Utilities,
Development*)
1376 Borregas Ave.
Sunnyvale, CA 94089-1004
T: 408-747-0440
F: 408-747-0688
B: 408-747-0621
URL: http://www.oritools.com/

Overland Data (*Peripherals, Tape*)
8975 Balboa Ave.
San Diego, CA 92123-1599
T: 619-571-5555
F: 619-571-0982
URL:
http://www.ovrland.com/~odisales/

Pacific Bell (*Communications,
Hardware*)
URL: http://www.pacbell.com/
(Includes detailed service information)

Figure 3-58 Panamax.

Pacific Data Products (*Computers, Printers*)
9855 Scranton Rd.
San Diego, CA 92121
T: 619-552-0880
F: 619-552-0889
B: 619-452-6329

Pacific Microelectronics (PMC)
(*Manufacturer, Custom*)
10575 S.W. Cascade Blvd.
Portland, OR 97223-9912
T: 503-684-5657
F: 503-620-8051
URL: http://www.pmcnet.com/~pmc/

Packard Bell (*Computers, Systems*)
31717 La Tienda Dr.
Westlake Village, CA 91362

T: 818-865-1555
F: 818-865-0379
B: 818-313-8601
CIS: go PACARDBELL
Prodigy: jump PACKARDBELL
URL: http://www.packardbell.com/

PAiA Electronics (*Distributor, Hobby*)
3200 Teakwood Ln.
Edmond, OK 73013
T: 405-340-6300
F: 405-340-6378
URL: http://www.paia.com/paia/

Panamax (*Equipment, Power Products*)
150 Mitchell Blvd.
San Rafael, CA 94903-2057
T: 415-499-3900
F: 415-472-5540

URL: http://www.hooked.net/
panamax/

**Panasonic Communications & Systems
Co.** (*Computers, Printers and Drives*)
2 Panasonic Way
Secaucus, NJ 07094
T: 201-348-7000
F: 201-348-8164
B: 201-863-7845
URL: http://www.panasonic.com/

Paradise (*Controllers, Video*)
T: 800-978-3029
B: 415-335-2744
E-mail: support@www.paradisemmp.com
URL: http://www.paradisemmp.com

Parallax (*SW Utilities, Development*)
3805 Atherton Rd., Suite 102
Rocklin, CA 95765
T: 916-624-8333
F: 916-624-8003
B: 916-624-7101
FaxBack: 916-624-1869
E-mail: info@parallaxinc.com
FTP: ftp://ftp.parallaxinc.com/
URL: http://www.parallaxinc.com/

PartNET (*Distributor, Components*)
505 Wakara Way, Research Park
Salt Lake City, UT 84108
T: 800-727-8061
F: 801-585-5238
URL: http://www.part.net:80/partnet/

PC Catalog (*Publisher, Computer*)
Part of PC Today *magazine*
URL: http://www.peed.com/
pccatalog.html

PC Computing (*Publisher, Computer*)
URL: http://www.zdnet.com/~pccomp/

PC-Kwik Corporation (*SW Utilities,
Business*)
3800 SW Cedar Hills Blvd., Suite 260
Beaverton, OR 97005
T: 503-644-5644
FaxBack: 503-646-8267
CIS: 76004,151

PCMCIA (*Services, Standards*)
1030G East Duane Ave.
Sunnyvale, CA 94086
T: 408-720-0107
F: 408-720-9416
B: 408-720-9388
E-mail: office@pcmcia.org
URL: http://www.pc-card.com/

PCs Compleat (*Computers, Systems*)
T: 508-624-6400
E-mail: sales@pcscompleat.com

PC Universe (*Distributor,
PC Assemblies*)
T: 407-447-0050
F: 407-447-7549
E-mail: sales@pcuniverse.com
URL: http://www.pcuniverse.com/

PC Warehouse (*Distributor,
PC Assemblies*)
T: 215-855-7898
URL: http://www.pcwarehouse.com/

Peachpit Press (*Publishers, Internet*)
2414 Sixth St.
Berkeley, CA 94710
T: 510-548-4393
F: 510-548-5991
URL: http://www.peachpit.com/

Figure 3-59 PC Universe.

Peak Computing   (*Computers, Systems*)
244 West Main St.
Goshen, NY 10924
T: 800-732-5908
F: 914-294-6287
URL: http://www.pccatalog.com/

Peerless Radio Corporation
(*Distributor, Components*)
URL: http://www.peerless-rc.com/

PerComp Microsystems, Inc.
(*Computers, Systems*)
T: 800-856-6688
F: 214-234-8658
B: 214-234-8376
URL: http://www.percomp.com/pc/

Performance Technology   (*Networking, Software*)
T: 210-979-2000
F: 210-979-2002
CIS: go PCVENH
E-mail: support@perftech.com
URL: http://www.perftech.com/

Peripherals Unlimited   (*Distributor, PC Assemblies*)
1500 Kansas Ave., Suite 4C
Longmont, CO 80501
T: 303-772-1482
F: 303-772-9430
URL: http://www.csn.net/peripherals/

Peripheral Technology Group, Inc.
(*Networking, Hardware and Software*)
7625 Golden Triangle Dr., Suite T
Eden Prairie, MN 55344
T: 612-942-7474
F: 612-942-7586
E-mail: sales@ptgs.com
URL: http://www.ptgs.com/

Persoft, Inc. (*Networking, Software*)
465 Science Dr.
P.O. Box 44953
Madison, WI 53744-4953
T: 608-273-6000
F: 608-273-8227
URL: http://www.persoft.com/

Phihong (*Equipment, Power Supplies*)
3005 Brodhead Rd.
Bethlehem, PA 18017
T: 610-866-9911
F: 610-866-9920
URL: http://www.phihongusa.com/

Phillips ECG (*Manufacturer, Components*)
URL: http://www.philips.com/
(Includes links to all business units)
URL: http://www.semiconductors.
philips.com/ps/
URL: http://server1.pa.hodes.com/ps/

Phoenix Technologies Ltd.
(*ICs, BIOS*)
846 University Ave.
Norwood, MA 02062-3950
T: 617-551-4000
F: 617-551-3750
URL: http://www.ptltd.com/

Picker International, Inc. (*Equipment, Imaging*)
595 Miner Rd.
Highland Heights, OH 44143
T: 216-473-3000
URL: http://www.picker.com/

Pinnacle Publishing, Inc. (*Publishers, Computer*)
P.O. Box 888
Kent, WA 98035
T: 206-251-1900
F: 206-251-5057
URL: http://www.pinpub.com/

Pioneer Electronics (*Distributor, Components*)
T: 216-498-6549
F: 216-498-6886
URL: http://www.pios.com/default.htm

Pipeline Network (*SW Utilities, Business*)
T: 212-267-3636
E-mail: support@pipeline.com

Pivar Computing Services, Inc.
(*Services, Data Conversion*)
165 Arlington Heights Rd.
Buffalo Grove, IL 60089
T: 708-459-6010
F: 708-459-6095
URL: http://www.pivar.com/

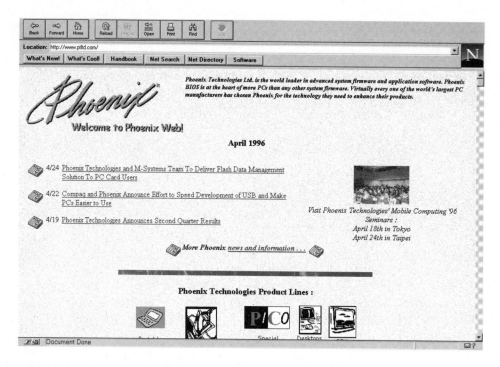

Figure 3-60 Phoenix Technologies Ltd.

PKWARE, Inc. (*SW Utilities, Data Compression*)
9025 N. Deerwood Dr.
Brown Deer, WI 53223-2437
T: 414-354-8699
F: 414-354-8559
B: 414-354-8670
E-mail: info@pkware.com
URL: http://www.pkware.com/

Plaintree Systems (*Networking, Hardware*)
Prospect Place, 9 Hillside Ave.
Waltham, MA 02154
T: 617-290-5800
F: 617-290-0963
URL: http://www.plaintree.com/plaintree/

URL: http://www.plaintree.on.ca/plaintree/

Plannet Crafters (*SW Utilities, Business*)
T: 770-998-8664
F: 770-998-8197
B: 770-740-8583
CIS: 73040,334
E-mail: Dmandell@aol.com

Plessey Semiconductors
(*Manufacturer, Components*)
URL: http://www.ge.com/
(Includes links to GE business units)

Figure 3-61 Power Express.

Plextor (*Computers, Drives*)
4255 Burton Dr.
Santa Clara, CA 95054
T: 408-980-1838
F: 408-986-1010
B: 408-986-1569

Portrait Display Labs (*Peripherals, Monitors*)
6665 Owens Dr.
Pleasanton, CA 94588
T: 510-227-2700
F: 510-227-2705
URL: http://www.sirius.com/~inform/

Powell Electronics, Inc. (*Distributor, Components*)
T: 800-235-7880

E-mail: info@powell.com
URL: http://www.powell.com/
(Includes contacts to nation-wide offices)

Power Convertibles (*Equipment, Power Products*)
3450 S. Broadmont Dr.
Tucson, AZ 85713
T: 520-628-8292
F: 520-770-9369
E-mail: sales@pccl.com
URL: http://www.pccl.com/

Power Express (*Computers, Batteries*)
T: 800-POWER-EX
URL: http://www.powerexpress.com/

PowerQuest   (*Diagnostics, Software*)
P.O. Box 275
Orem, UT 84059-0275
T: 801-221-9842
F: 801-226-2519
E-mail: barlow@itsnet.com
URL: http://www.powerquest.com/

Powersoft   (*SW Utilities, Development*)
561 Virginia Rd.
Concord, MA 01742-2732
T: 508-287-1500
F: 508-287-1600
B: 508-287-1850
E-mail: sales@powersoft.com
URL: http://www.powersoft.com/

Power Trends   (*Equipment, Power Products*)
1101 North Raddant Rd.
Batavia, IL 60510
T: 708-406-0900
URL: http://www.powertrends.com/isr/

PMC-Sierra   (*ICs, Communication*)
8501 Commerce Ct.
Burnaby, BC, V5A 4N3, Canada
T: 604-668-7300
URL: http://www.pmc-sierra.com/

Practical Peripherals, Inc.
(*Manufacturer, Modems*)
375 Conejo Ridge Ave.
Thousand Oaks, CA 91361
T: 805-497-4774
F: 805-374-7200
Tech Support: 805-496-7707
Tech Fax: 805-374-7216
B: 805-496-4445

Precision Digital Images   (*Equipment, Imaging*)
8520 154th Ave. NE
Redmond, WA 98052
T: 206-882-0218
F: 206-867-9177
URL: http://www.precisionimages.com/

Processor Magazine   (*Publisher, Electronics*)
PO Box 85518
Lincoln, NE 68501
T: 800-247-4880
F: 402-479-2120
B: 402-477-2283
URL: http://www.peed.com/

Prodigy   (*Services, On-line*)
URL: http://www.prodigy.com/

Promise Technology   (*Controllers, Drive*)
B: 408-452-1267
URL: http://www.promise.com/

Protel Technology   (*Simulation, Circuit CAD*)
4675 Stevens Creek Blvd. Suite 200
Santa Clara, CA 95051
T: 408-243-8143
F: 408-243-8544
E-mail: salesusa@protel.com
URL: http://www.protel.com/

Proteon   (*Networking, Hardware*)
T: 508-898-2800
E-mail: pro@proteon.com
FTP: ftp://ftp.proteon.com
URL: http://www.proteon.com/

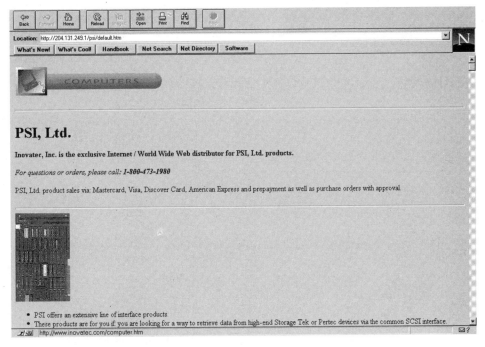

Figure 3-62   PSI-Squared, Ltd.

Provantage Corporation   (*Distributor, Software*)
7249 Whipple Ave., NW
North Canton, OH 44720-7143
T: 330-494-3781
F: 330-494-5260
URL: http://www.provantage.com/

PSI-Squared, Ltd.   (*Controllers, Drive*)
Inovatec, Inc.
T: 719-596-9360
F: 719-596-9384
E-mail: inovatec@inovatec.com
URL: http://204.131.249.1/psi/

Public Software Library   (*Distributor, Software*)
PO Box 35705
Houston, TX 77235

T: 713-524-6394
F: 713-524-6398
B: 713-442-6704
E-mail: nelson.ford@pslonline.com

Qlogic Corporation   (*Controllers, SCSI*)
3545 Harbor Blvd., PO Box 5001
Costa Mesa, CA 92628-5001
T: 714-438-2200
B: 714-708-3170
URL: http://www.qlc.com/

QMS   (*Computers, Printers*)
One Magnum Pass
Mobile, AL 36618
T: 334-633-4300
URL: http://www.qms.com/

QNX Software Systems, Ltd.
(*SW Utilities, Development*)
175 Terence Matthews Crescent
Kanata, Ontario K2M 1W8 Canada
T: 613-591-0931
F: 613-591-3579
E-mail: info@qnx.com
URL: http://www.qnx.com/

Q-Sound (*Peripherals, Sound*)
2748-37th Ave. N.E.
Calgary, Alberta, T1Y 5L3 Canada
T: 403-291-2492
F: 403-250-1521
CIS: go QSOUND
E-mail: tech.support@qsound.com
URL: http://www.qsound.com/

Qualcomm (*SW Utilities, Communication*)
6455 Lusk Blvd.
San Diego, CA 92121-2779
T: 619-587-1121
F: 619-658-2100
URL: http://lorien.qualcomm.com/
URL: http://www.qualcomm.com/

Quadram (*Controllers, Drive*)
9 Mecca Way
Norcross, GA 30093
T: 404-923-6666
B: 404-564-5678

Qualitas, Inc. (*SW Utilities, Memory Management*)
7101 Wisconsin Ave., #1386
Bethesda, ND 20814
T: 301-907-6700
F: 301-718-6060
B: 301-907-8030
CIS: 75300,1107
AOL: keyword QUALITAS

Quality Semiconductor (*ICs, Logic*)
851 Martin Ave.
Santa Clara, CA 95050
T: 408-450-8000
F: 408-496-0773
URL: http://www.qualitysemi.com/

Quanta Micro Corporation
(*Distributor, PC Assemblies*)
T: 310-907-1180
F: 310-907-1186
URL: http://www.q.com/

Quantex Microsystems, Inc.
(*Computers, Systems*)
400B Pierce St.
Somerset, NJ 08873
T: 800-836-0566
URL: http://www.quantex.com/

Quantum Corporation (*Computers, Drives*)
500 McCarthy Blvd.
Milpitas, CA 95035
T: 408-894-4000
F: 408-894-3282
B: 408-434-1664
URL: http://www.quantum.com/

Quarterdeck Corporation (*SW Utilities, Memory Management and Internet*)
150 Pico Blvd.
Santa Monica, CA 90405
T: 310-392-9701
F: 310-314-3217
B: 310-314-3227
CIS: go QUARTERDECK
E-mail: support@qdeck.com
E-mail: info@qdeck.com
FTP: ftp.qdeck.com
URL: http://www.qdeck.com/

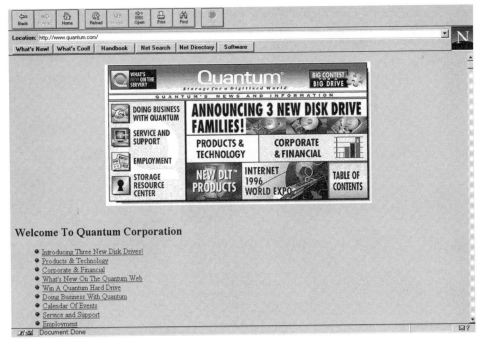

Figure 3-63 Quantum Corporation.

Quay Computers (*Computers, Systems*)
Bowen House, 1 Bowen St.
Wellington, New Zealand
T: 644-499-3368
F: 644-499-3327
E-mail: quayw@actrix.gen.nz
URL: http://www.quay.co.nz/

Quickturn (*Manufacturer, ICs*)
440 Clyde Ave.
Mountain View, CA 94043
T: 415-967-3300
F: 415-967-3199
URL: http://www.quickturn.com/

Racal-Datacom Corporation
(*Networking, Hardware*)
1601 North Harrison Parkway
PO Box 407044
Fort Lauderdale, FL 33340-7044
T: 954-846-1601
F: 954-846-4942
URL: http://www.racal.com/racal.html

Radio Control Systems, Inc.
(*Equipment, Communication*)
8125 Ronson Rd., Suite G
San Diego, CA 92111
T: 619-560-7008
F: 619-292-6955
E-mail: rcsi@cts.com
URL: http://www.cts.com/browse/rcsi/

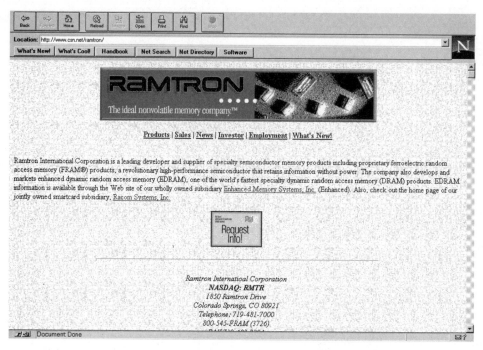

Figure 3-64 Ramtron International.

RadiSys (*Computers, Dedicated*)
15025 SW Koll Parkway
Beaverton, OR 97006-6041
T: 503-646-1800
F: 503-646-1850
E-mail: info@radisys.com
URL: http://www.radisys.com/

Radius (*SW Utilities, Business*)
215 Moffett Park Dr.
Sunnyvale, CA 94089-1374
T: 408-541-5700
F: 408-541-5008
B: 408-541-6190
CIS: go MACBVEN
AOL: keyword RADIUS
URL: http://www.radius.com/

Rainbow Technologies (*Equipment, Dongles*)
50 Technology Dr.
Irvine, CA 92718-2301
T: 714-450-7300
F: 714-450-7450
URL: http://www.rnbo.com/

Rambus (*ICs, Logic*)
T: 415-903-3800
URL: http://www.rambus.com/

Ramtron International (*ICs, Logic*)
1850 Ramtron Dr.
Colorado Springs, CO 80921
T: 719-481-7000
F: 719-481-9294
URL: http://www.csn.net/ramtron/

Rancho Technology, Inc. (*Controllers, SCSI*)
10783 Bell Ct., #109
Rancho Cucamonga, CA 91730
T: 909-987-3966
F: 909-989-2365
B: 909-980-7699
E-mail: rancho@realm.com

Raritan Computer, Inc. (*Equipment, Printer Sharing*)
Building 10, Suite No. 1
Ilene Court
Belle Mead, NJ 08502
T: 908-874-4072
F: 908-874-5274
E-mail: sales@raritan.com
URL: http://www.raritan.com/

RDI Computer (*Computers, Workstations*)
2300 Faraday Ave.
Carlsbad, CA 92008
T: 800-RDI-LITE
F: 619-931-1063
URL: http://www.rdi.com/home.html

Realtek (*Controllers, Drive*)
1F, No 11 Industry East Rd. IX
Science-Based Industrial Park
Hsinchu, Taiwan, ROC
T: 886-35-780211
F: 886-35-776047
E-mail: chtseng@realtek.com.tw
URL: http://www.realtek.com/

Reptron Electronics (*Distributor, Components*)
14401 McCormick Dr.
Tampa, FL 33626-3046
T: 813-854-2351

F: 813-855-0942
URL: http://www.reptron.com/

Republic Electronics (*Manufacturer, Components*)
476 Blackman St.
Wilkes-Barre, PA 18702-6008
T: 717-823-9900
F: 717-825-6412
URL:
http://www.microserve.net/~republic/

Reynolds Data Recovery (*Diagnostics, Software*)
13205 North 87th
Longmont, CO 80501
T: 303-776-7110
F: 303-776-7277
URL: http://www.data-recovery.com/reynolds/

RicheyCypress Electronics (*Distributor, Components*)
7441 Lincoln Way
Garden Grove, CA 92641
T: 714-898-8288
F: 714-897-7887
URL: http://www.richeyelec.com/

Ricoh Company, Ltd. (*Computers, Printers*)
5 Dedrick Place, West
Caldwell, New Jersey 07006
T: 201-882-2000
Printers: 800-955-3453
Scanners: 800-955-3453
URL:
http://www.ricoh.co.jp/index_e.html

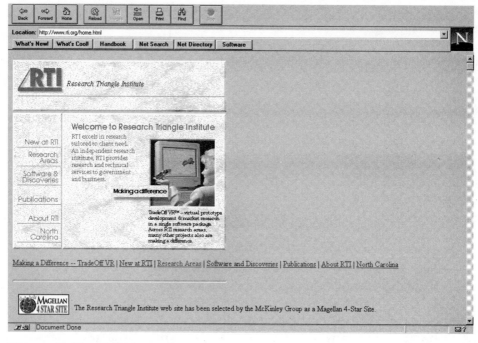

Figure 3-65   RTI.

Rockwell International   (*ICs, Communication*)
4311 Jamboree Rd.
Newport Beach, CA 92660-3095
T: 714-833-4600
URL: http://www.nb.rockwell.com/
URL: http://www.rockwell.com/
(Includes links to all business units)

Ross Technology, Inc.   (*ICs, SPARC Processors*)
5316 Hwy. 290 West
Austin, TX 78735
T: 512-349-3108
F: 512-349-3101
URL: http://www.ross.com/

RTI (Research Triangle Institute)
(*Services, Research*)
3040 Cornwallis Rd.
Research Triangle Park, NC 27709
T: 919-541-6000
URL: http://www.rti.org/home.html

Sager Electronics   (*Distributor, Components*)
60 Research Rd.
Hingham, MA 02043
T: 800-724-3780
F: 800-268-8001
E-mail: sagerinfo@sager.com
URL: http://www.sager.com/

SAIC   (*Services, Research*)
10260 Campus Point Dr.
San Diego, CA 92121

T: 619-546-6000
URL: http://www.saic.com/

Samsung Electronics America
(*Computers, Systems*)
T: 800-726-7864
FaxBack: 800-229-2239
B: 201-691-6238
URL: http://www.samsung.com/

Samtec, Inc. (*Computers, Connectors*)
PO Box 1147
810 Progress Blvd.
New Albany, IN 47151-1147
T: 812-944-6733
F: 812-948-5047
URL: http://www.samtec.com/

SanDisk Corporation (*Computers, Flash Storage*)
3270 Jay St.
Santa Clara, CA 95054
T: 408-562-0595
F: 408-562-3403
E-mail: info@sandisk.com
URL: http://www.sandisk.com/

Sanyo (*Computers, Batteries*)
URL: http://www.sanyo.co.jp/koho/e_index.html

SaRonix (*Manufacturer, Components*)
151 Laura Ln.
Palo Alto CA 94303
T: 415-856-6900
F: 415-856-4732
E-mail: saronix@connectinc.com
URL: http://www.saronix.com/

Sattel (*Communications, Hardware*)
9145 Deering Ave.
Chatsworth, CA 91311

T: 818-718-6437
F: 818-718-1847
URL: http://www.sattel.com/

SB Electronics, Inc. (*Manufacturer, Components*)
131 South Main St.
Barre, VT 05641-4854
T: 802-476-4146
F: 802-476-4149
E-mail: SBE@InternetMCI.com
URL: http://www.to-be.com/sbe/

SBE, Inc. (*Networking, Hardware*)
4550 Norris Canyon Rd.
San Ramon, CA 94583-1369
T: 800-214-4723
F: 510-355-2042
B: 510-355-2048
E-mail: support@sbei.com
URL: http://www.sbei.com/

SBI Computer Products (*Distributor, PC Assemblies*)
128 Atlantic Ave.
Toronto, Ontario, M6K 1X9 Canada
T: 416-537-2611
F: 416-537-1484
URL: http://www.the-wire.com/SBI/

Schaffner EMC (*Equipment, EMI*)
9B Fadem Rd.
Springfield, NJ 07081
T: 201-379-7778
F: 201-379-1151
URL: http://www.schaffner.com/

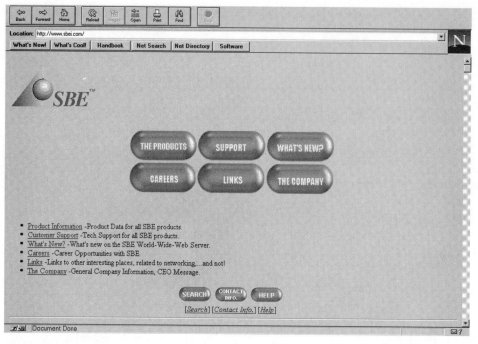

Figure 3-66 SBE, Inc.

Schurter (*Manufacturer, Components*)
Schurter, Inc. 1016 Clegg Ct.
PO Box 750158
Petaluma, CA 94975
T: 707-778-6311
F: 707-778-6401
URL: http://www.schurterinc.com/

SCii Telecom (*Communications, ISDN*)
5 Ter Rue Du Dome
75116 Paris, France
T: 331-4417-4417
F: 331-4417-4419
E-mail: technical@scii.co.uk
URL: http://www.scii.co.uk/

Scope Systems (*Services, Repair*)
76 Cranbrook Rd., Suite #203
Hunt Valley, MD 21030

T: 410-527-1090
F: 410-527-0957
E-mail: Scope@ScopeSys.com
URL: http://www.scopesys.com/

Seagate Technology (*Computers, Drives*)
920 Disc Dr.
Scotts Valley, CA 95066
T: 408-438-6550
F: 408-429-6356
B: 408-438-8771
FaxBack: 408-438-2620
CIS: go SEAGATE
FTP: ftp://ftp.seagate.com/
techsupport/at/
URL: http://www.seagate.com/

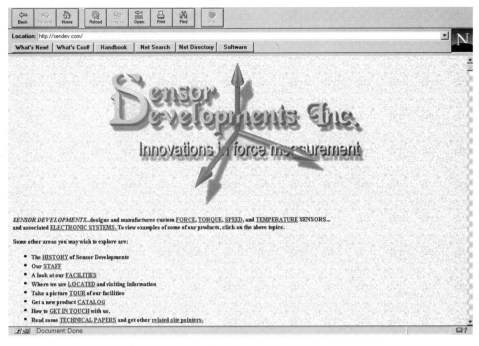

Figure 3-67   Sensor Developments, Inc.

Seeq Technology   (*ICs, Communication*)
47200 Bayside Pkwy.
Fremont, CA 94538
T: 510-226-7400
F: 510-657-2837
URL: http://www.seeq.com/

SEI   (*Distributor, Components*)
Sonepar Electronique International S.A.
2, Rue De La Tour-des-Dames
F-75009 Paris, France
T: 31-76-5722400
F: 31-76-5877743
E-mail: comm@sei.nl
URL: http://www.sei-europe.com:8008/

Sensor Development, Inc.
(*Manufacturer, Sensors*)
PO Box 290
Lake Orion, MI 48361-0290
T: 810-391-3000
F: 810-391-0107
CIS: 74720,3227
E-mail: sensor@sendev.com
URL: http://sendev.com/

Sequent Computer Systems, Inc.
(*Computers, Servers*)
15450 S.W. Koll Parkway
Beaverton, OR 97006-6063
T: 503-626-5700
URL: http://www.sequent.com/

Sermax Corporation (*Distributor, SIMM Adapters*)
207 E 94th St.
New York, NY 10128
T: 212-410-1597
F: 212-410-0452
URL: http://www.simmstack.com/

ServiceWare Corporation
(*SW Utilities, Business*)
2280 St. Laurent Blvd., Suite 104
Ottawa, Ontario, K1G 4K1 Canada
T: 613-521-7391
T: 613-521-5595
B: 613-521-2105
E-mail: netlink@service.magi.com

SGS-Thompson Microelectronics
(*Manufacturer, ICs*)
55 Old Bedford Rd.
Lincoln, MA 01773
T: 617-259-0300
F: 602-867-6102
URL: http://www.st.com/
(Includes links to worldwide distributors)

Shallco, Inc. (*Manufacturer, Components*)
PO Box 1089
Smithfield, NC 27577
T: 919-934-3135
F: 919-934-3298
E-mail: pdorman@nando.net
URL: http://www.nando.net/shallco/

Sharp Electronics (*Peripherals, LCDs*)
Sharp Plaza
Mahwah, NJ 07430-2135
T: 201-529-8200
F: 201-529-9636
URL: http://www.sei-europe.com:8008/
sei/shr/shrmenu.htm

Shiva Corporation (*Networking, Hardware and Software*)
28 Crosby Dr.
Bedford, MA 01730-1437
T: 617-270-8300
F: 617-270-8599
B: 617-273-0023
CIS: go SHIVA
URL: http://www.shiva.com/

Shop Direct, Inc. (*Computers, Systems*)
12501 Philadelphia St.
Whittier, CA 90601
T: 800-215-7476
F: 800-316-7467
E-mail: shopdirect@aol.com

Siemens USA (*Manufacturer, ICs*)
1301 Avenue of the Americas
New York, NY 10019
T: 212-258-4000
F: 212-767-0580
URL: http://www.siemens.com/

Silicom Electronics, Inc. (*Computers, Multimedia*)
3335 Kifer Rd.
Santa Clara, CA 95051
T: 408-774-0700
F: 408-774-0733
URL: http://www.best.com/~frankj/
silicom.htm

Silicon Engineering (*Manufacturer, ICs*)
269 Mt. Hermon Rd., Suite 101
Scotts Valley, CA 95066
T: 408-438-5330
F: 408-438-8509
E-mail: info@sei.com
URL: http://www.sei.com/

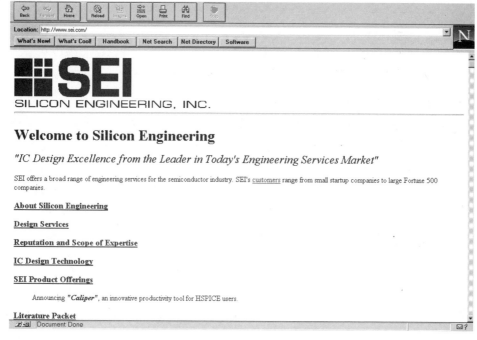

Figure 3-68 Silicon Engineering.

Silicon Graphics (*Computers, Workstations*)
URL: http://www.sgi.com/
(Includes links to worldwide sales offices)

Silicon Soft (*Peripherals, Data Acquisition*)
5131 Moorpark Ave. #303
San Jose, CA 95129
T: 408-446-4521
F: 408-445-5196
URL: http://www.siliconsoft.com/

Silicon Systems, Inc. (*ICs, Mixed-Signal*)
14351 Myford Rd.
Tustin, CA 92680-7022
T: 714-573-6000
F: 714-573-6914

E-mail: info@ssi1.com
URL: http://www.ssi1.com/

Silicon Valley Research (*Controllers, Drive*)
300 Ferguson, Bldg. 300
Mountain Valley, CA 94043
T: 415-962-3000
URL: http://www.svri.com/

SIMM Saver Technology, Inc.
(*Distributor, SIMM Adapters*)
1820 E. 1st
Wichita, KS 67214
T: 316-264-2244
F: 316-264-4445
E-mail: simmsave@dtc.net
URL: http://wwww.solgate.com~sti/

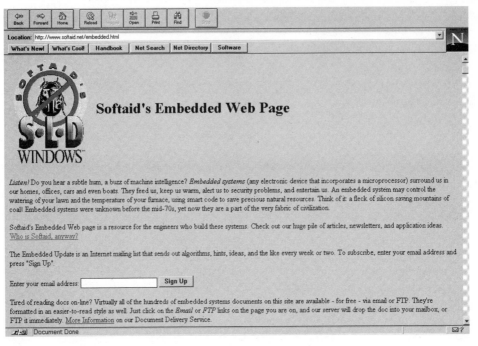

Figure 3-69   Softaid, Inc.

Simple Technology   (*Distributor, Memory*)
3001 Daimler St.
Santa Ana, CA 92705
T: 714-476-1180
F: 714-476-1209
B: 714-476-9034
URL: http://www.simpletech.com/

Simtek Corporation   (*Distributor, Memory*)
T: 719-531-9444
E-mail: info@simtek.com
URL: http://www.csn.net/simtek/

Skyline Technology   (*Networking, Hardware*)
T: 415-377-0199
F: 415-377-0599

URL: http://www.skylinetech.com/
(Includes distributor contacts)

Softaid, Inc.   (*Equipment, Emulation*)
8310 Guilford Rd.
Columbia, MD
T: 410-290-7760
F: 410-381-3253
E-mail: emulate@softaid.net
URL: http://www.softaid.net/
embedded.html

SoftInfo   (*SW Utilities, Reference*)
823 E. Westfield Blvd.
Indianapolis, IN 46220
T: 317-251-7727
F: 317-251-7813
E-mail: softinfo@icp.com
URL: http://www.icp.com/softinfo/
homepg_3.html

SoftNet Systems (*Equipment, Imaging*)
717 Forest Ave.
Lake Forest, IL 60045
E-mail: info@Softnet.com
URL:
http://www.softnet.com/index.html

Software Publisher's Association (*Services, Anti-Piracy*)
1730 M St., NW, Suite 700
Washington, DC 20036-4510
T: 202-452-1600
F: 202-223-8756
URL: http://www.spa.org/

Software Publishing Corporation (SPC) (*SW Utilities, Business*)
111 N. Market St.
San Jose, CA 95113
T: 408-537-3000
F: 408-980-1518
CIS: go SPC
URL: http://www.spco.com/

S-MOS Systems (*Manufacturer, ICs*)
2460 North First St.
San Jose, CA 95131-1002
T: 408-922-0200
F: 408-922-0238
URL: http://www.smos.com/

Sony Electronics (*Computers, Multimedia and Drives*)
Sony Dr.
Park Ridge, NJ 07656
T: 408-894-0555
FaxBack: 408-955-5505
B: 408-955-5107
URL: http://www.sel.sony.com/
URL: http://www.sony.com/

Southwire (*Manufacturer, Materials*)
One Southwire Dr.
Carrollton, GA 30119
T: 770-832-4242
URL: http://www.southwire.com/

Spatializer Audio Labs (*Peripherals, Sound*)
20700 Ventura Blvd., Suite 134
Woodland Hills, CA 91364-2357
T: 818-227-3370
F: 818-227-9750
URL: http://www.spatializer.com/

Sparc International (*Services, Information*)
535 Middlefield Rd., Suite 210
Menlo Park, CA 94025
T: 415-321-8692
F: 415-321-8015
E-mail: webinfo@sparc.com
URL: http://www.sparc.com/

Spectron (*SW Utilities, Development*)
315 Bollay Dr.
Santa Barbara, CA 93117
T: 805-968-5100
F: 805-968-9770
URL: http://spectron.internet-cafe.com/

Spectrum Trading (*Distributor, PC Assemblies*)
2511 Garden Rd., Building B, Suite 100
Monterey, CA 93940
T: 408-655-4000
F: 408-655-4019
URL: http://www.spectrum-t.com/

Figure 3-70   Spatializer Audio Labs.

Sprague Magnetics, Inc.   (*Services, Repair*)
15720 Stagg St.
Van Nuys, CA 91406
T: 818-994-6602
F: 818-994-2153
URL: http://home.earthlink.net/~sprague-magnetics/
URL: http://home.earthlink.net/~sprague-magnetics/qicpage.html

SPRY   (*Services, On-line*)
316 Occidental Ave. S
Seattle, WA 98104
E-mail: winmbox63@spry.com
URL: http://www.spry.com/

Spyglass   (*SW Utilities, Internet*)
1240 E. Diehl Rd.
Naperville, IL 60563
T: 708-505-1010
F: 708-505-4944
URL: http://www.spyglass.com/

SRS Labs   (*Peripherals, Sound*)
2909 Daimler St.
Santa Ana, CA 92705
T: 714-442-1070
F: 714-852-1099
URL: http://www.srslabs.com/

Stac Electronics   (*SW Utilities, Data Compression*)
5993 Avenida Encinas
Carlsbad, CA 92008
T: 619-431-7474
F: 619-431-0880

Figure 3-71 Standard Microsystems Corporation.

B: 619-431-5956
CIS: go STAC
AOL: keyword STAC

Standard Microsystems Corporation (SMC) (*Networking, Hardware/ICs, Communication*)
35 Marcus Blvd.
Hauppauge, NY 11788
T: 516-273-3100
F: 516-273-5550
B: 516-434-3162
URL: http://www.smc.com/

Standard Printed Circuits, Inc.
(*Manufacturer, PCB*)
44 South Main St.
Sherburne, NY 13460
T: 800-555-0980
F: 607-674-4409

URL: http://www.halcyon.com/ rbormann/spc.htm

Starfish (*SW Utilities, Internet*)
1700 Green Hills Rd.
Scotts Valley, CA 95066
T: 800-765-7839
CIS: go STARFISH
AOL: keyword STARFISH
URL: http://www.starfishsoftware.com/

Star Micronics (*Computers, Printers*)
70-D Ethel Rd. W.
Piscataway, NJ 08854
T: 908-572-5550
F: 908-572-3129
B: 908-572-5010

STB Systems (*Controllers, Video*)
1651 N. Glenville
Richardson, TX 75085
T: 214-234-8750
F: 214-234-1306
B: 214-437-9615
CIS: go STBSYS
E-mail: support@stb.com
URL: http://www.stb.com/

Sterling Instruments (*Distributor, Electro-Mechanical Components*)
2101 Jericho Turnpike
Box 5416
New Hyde Park, NY 11042-5416
T: 516-328-3300
F: 516-326-8827
E-mail: catalog@sdp-si.com
URL: http://www.sdp-si.com/

Stone Technologies (*Security, Hardware*)
#4 Stone Court - 2311 Westrock Dr.
Austin, TX 78704-5837
T: 512-440-1234
E-mail: stc@io.com
URL: http://www.webcom.com/~stc/

Storage USA (*Distributor, Drives*)
A division of New MMI Corporation
2400 Reach Rd.
Williamsport, PA 17701
T: 717-327-9200
F: 717-327-1217
URL: http://www.storageusa.com/

Stylus Product Group (*SW Utilities, Development*)
201 Broadway
Cambridge, MA 02139
T: 617-621-9545
F: 617-621-7862

E-mail: info@stylus.com
URL: http://www.stylus.com/

Summit Distributors (*Distributor, Components*)
916 Main St.
Buffalo, NY 14202
T: 800-6-SUMMIT
F: 800-678-2866
E-mail: summitdist@aol.com
URL: http://members.aol.com/SummitDist/

Sumo (*Peripherals, Sound*)
5312 Derry Ave., Suite K
Agoura Hills, CA 91301
T: 818-706-9974
F: 818-706-1139
URL: http://emporium.turnpike.net/S/sumo/index.htm

Sun Microsystems (*Computers, Workstations*)
2550 Garcia Ave.
Mountain View, CA 94043-1100
T: 214-788-3188
URL: http://www.sun.com/

Sunshine Computers (*Distributor, PC Assemblies*)
1240 East Newport Center Dr.
Deerfield Beach, FL 33442
T: 800-828-2992
F: 954-422-9608
E-mail: ginog@gate.net
URL: http://www.sunshinec.com/

Supra Communications (*Computers, Modems*)
A division of Diamond Multimedia Systems
312 SE Stonemill Dr., Suite 150
Vancouver, WA 98684

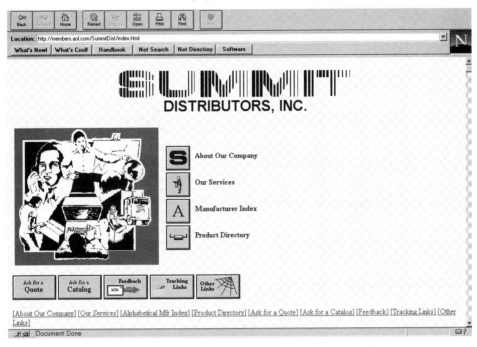

Location: http://members.aol.com/SummitDist/index.html

About Our Company

Our Services

Manufacturer Index

Product Directory

Ask for a Quote | Ask for a Catalog | Feedback | Tracking Links | Other Links

[About Our Company] [Our Services] [Alphabetical Mfr Index] [Product Directory] [Ask for a Quote] [Ask for a Catalog] [Feedback] [Tracking Links] [Other Links]

Document Done

Figure 3-72   Summit Distributors.

T: 360-604-1400
F: 360-604-1401
B: 541-967-2444
CIS: go SUPRA
FTP: ftp.supra.com
URL: http://www.supra.com/

Surplus Software, Inc.   (*Distributor, PC Assemblies and Software*)
489 North Eight
Hood River, OR 97031
T: 800-753-7877
E-mail: granny@surplusdirect.com
URL: http://surplusdirect.com/

SVEC Computer Corporation
(*Networking, Hardware*)
2691 Richter Ave., Suite 130
Irvine, CA 92714
T: 714-756-2233

F: 714-756-1340
E-mail: info@SVEC.com
URL: http://www.svec.com/

Swan Technologies Corporation
(*Computers, Systems*)
313 Boston Post Rd. West, Suite 200
Marlboro, MA 01752
T: 508-460-1977
F: 508-480-0156
URL: http://www.swantech.com/

Sweetwater Sound, Inc.   (*Peripherals, Sound*)
5335 Bass Rd.
Fort Wayne, IN 46808
T: 219-432-8176
F: 219-432-1758
E-mail: funmidi@sweetwater.com
URL: http://www.sweetwater.com/

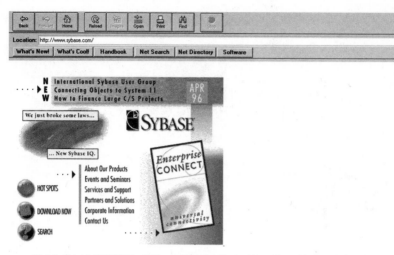

Figure 3-73 Sybase.

Sybase (*Networking, Software*)
6475 Christie Ave.
Emeryville, CA 94608-1050
T: 510-922-3555
URL: http://www.sybase.com/

Symantec Corporation (*Diagnostics, Virus*)
10201 Torre Ave.
Cupertino, CA 95014-2132
T: 408-253-9600
F: 408-255-9341
B: 408-973-9598
URL: http://www.symantec.com/

Symbol Technologies (*Equipment, Laser Scanning*)
One Symbol Plaza
Holtsville, NY 11742-1300

T: 516-738-2400
F: 516-738-2831
URL: http://www.symbol.com/

Synergy Semiconductor (*ICs, Mixed Signal*)
3450 Central Expressway
Santa Clara, CA 95051
T: 408-730-1313
F: 408-737-0831
E-mail: info@synergysemi.com
URL: http://www.synergysemi.com/

Synopsys (*Simulation, Circuit CAD*)
700 East Middlefield Rd.
Mountain View, CA 94043
T: 415-962-5000
URL: http://www.synopsys.com/

Figure 3-74 Tagram Systems Corporation.

SyQuest Technology, Inc. (*Computers, Drives*)
47071 Bayside Pkwy.
Fremont, CA 94538
T: 800-245-2278
F: 510-226-4102
B: 510-656-0473
E-mail: support-pc@syquest.com
FTP: ftp://ftp.syquest.com
URL: http://www.syquest.com/

SystemWare (*Services, Research*)
705 Archer Court
Herndon, VA 22070-5440
T: 703-318-0350
F: 703-318-8533
URL:
http://sysware.com/home/index.html

Tadiran (*Equipment, Communication*)
E-mail: info@telecomm.tadiran.co.il
URL:
http://www.telecomm.tadiran.co.il/

Tagram Systems Corporation
(*Computers, Systems*)
1451-B Edinger Ave.
Tustin, CA 92680
T: 714-258-3222
F: 714-258-3220
URL: http://www.tagram.com/

Taligent (*SW Utilities, Development*)
10355 N. De Anza Blvd.
Cupertino, CA 95014-2233
T: 408-255-2525
F: 408-777-5533
URL: http://www.taligent.com/

Talking Technologies (*Computers, Telephony*)
1125 Atlantic Ave.
Alameda, CA 94501
T: 510-522-3800
F: 510-522-5556
URL: http://www.tti.net/

Tandem Computers (*Computers, Servers*)
19333 Vallco Parkway
Cupertino, CA 95014-2599
T: 408-285-6000
F: 408-285-8611
E-mail: support@tandem.com
URL: http://www.tandem.com/

Tandy Corporation (*Computers, Systems*)
1800 One Tandy Center
Ft. Worth, TX 76102
T: 817-390-3700
F: 817-390-2647
URL: http://www.tandy.com/

Tangent Computer (*Computers, Servers*)
197 Airport Blvd.
Burlingame, CA 94010
T: 800-974-6658
F: 415-342-9380
URL: http://www.tangent.com/

Tanisys Technology (*Peripherals, Input Devices*)
1310 RR 620 South, Suite B195
Austin, TX 78758
T: 512-263-1700
F: 512-263-1763
URL: http://www.tanisys.com/

TAPEDISK Corporation (*Peripherals, Tape*)
200 Main St., Ste 210
Menomonie, WI 54751
T: 715-235-3388
F: 715-235-3818
CIS: 73174,2464

Tauber Electronics (*Computers, Batteries*)
10656 Roselle St.
San Diego, CA 92121
T: 619-450-0088
F: 619-450-1410
URL: http://www.electriciti.com/
~batteri/

TC Computers (*Distributor, PC Assemblies*)
T: 504-733-2527
E-mail: tech@tccomputers.com
URL: http://www.tccomputers.com/

Teac America, Inc. (*Computers, Drives*)
7733 Telegraph Rd.
Montbello, CA 90640
T: 213-726-0303
F: 213-727-7656
B: 213-727-7660

Team American (*Distributor, PC Assemblies*)
16104 Covello St.
Van Nuys, CA 91406

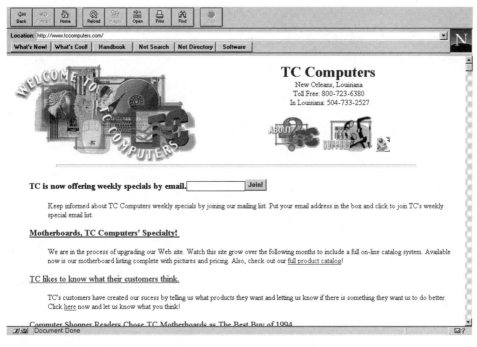

Figure 3-75 TC Computers.

T: 818-787-1920
F: 818-787-9118
E-mail: teamamer@ix.netcom.com
URL:
http://www.vir.com/JAM/team.html

Tech Data Corporation (*Distributor,*
PC Assemblies)
5350 Tech Data Dr.
Clearwater, FL 34620
T: 813-539-7429
F: 813-538-7876
B: 813-538-7090

Techlock Distributing (*Distributor,*
Components)
270 Regency Ridge, Suite 206
Dayton, OH 45459

T: 513-434-5078
F: 513-434-5079
URL: http://www.erinet.com/kenny/

TechTools, Inc. (*SW Utilities,*
Development)
3-I Taggart Dr.
Nashua, NH 03060
T: 603-888-8400
E-mail: info@techtools.com
URL: http://www.techtools.com/

TechWorks (*Distributor, Memory*)
4030 West Braker Ln.
Austin, TX 78759-5319
T: 512-794-8533
F: 512-794-8520
URL: http://www.techwrks.com/

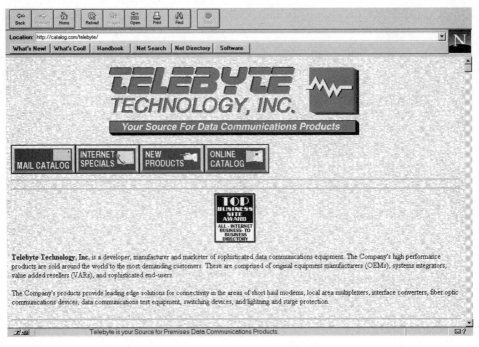

Figure 3-76 Telebyte Technology, Inc.

Tecmar, Inc. (*Controllers, Drive*)
6225 Cochran Rd.
Solon, OH 44139
T: 216-349-0600
F: 216-349-0851
B: 216-349-0853

Tekram Technology (*Controllers, Drive and Video*)
11500 Metric Blvd., #190
Austin, TX 78758
T: 512-833-6550
URL: http://www.tekram.com/

Tektronix (*Computers, Printers*)
URL: http://www.tek.com/
URL: http://www.tektronix.com/
(Includes links to all business units)

Teldor (*Manufacturer, Materials*)
Ein Dor, 19335, Israel
T: +972-6-770555
F: +972-6-770650
E-mail: teldor@netvision.net.il
URL: http://www.elron.net/teldor/

Telebit (*Computers, Modems*)
One Executive Drive
Chelmsford, MA 01824
T: 508-441-2181
F: 508-441-9060
URL: http://www.telebit.com/

Telebyte Technology, Inc. (*Computers, Modems*)
270 Pulaski Rd.
Greenlawn, NY 11740

T: 800-835-3298
F: 516-385-8184
E-mail: sales@telebyteusa.com
URL: http://catalog.com/telebyte/

Telecommunications Techniques Corp.
(*Diagnostics, Hardware*)
20400 Observation Dr.
Germantown, MD 20876
T: 301-353-1550
F: 301-353-0234
URL: http://www.ttc.com/

Telenex Corporation (*Networking, Hardware*)
General Signal, 13000 Midlantic Dr.
Mount Laurel, NJ 08054
T: 609-234-7900
F: 609-778-8700
FTP: ftp://ftp.gsnetworks.com
URL: http://www.gsnetworks.com/
mount_laurel.html

Tellurex (*Thermal Products, Heat Sinks*)
1248 Hastings St.
Traverse City, MI 49686
T: 616-947-0110
F: 616-947-5821
E-mail: tellurex@tellurex.com
URL:
http://tellurex.com/tellurex/home/

Teltone (*ICs, Communication*)
22121-20th Ave. SE
Bothell, WA 98021-4408
T: 206-487-1515
F: 206-487-2288
E-mail: info@teltone.com
URL: http://www.teltone.com/

TEM Computers (*Distributor, Memory*)
AAA Division, 7/16 Main St.
Blacktown, NSW, Australia 2148
T: +61-2-831-6792
F: +61-2-831-1855
E-mail: tem@geko.com.au
URL: http://www.magna.com.au/~tem/

TEMIC (Siliconix) (*Manufacturer, Components*)
2201 Laurelwood Rd.
Santa Clara, CA 95056-0951
T: 408-988-8000
F: 408-970-3950
URL: http://www.temic.de/

Terabyte (*Distributor, PC Assemblies*)
PO Box 2355
Boone, NC 28607-2355
T: 704-265-1234
E-mail: bbergin@terabyte.net
URL: http://www.terabyte.net/

Tera Computer Company (*Computers, Supercomputers*)
2815 Eastlake Ave. East
Seattle, WA 98102
T: 206-325-0800
F: 206-325-2433
E-mail: sales@tera.com
URL: http://www.tera.com/

Teradyne (*Diagnostics, Hardware*)
321 Harrison Ave.
Boston, MA 02118
T: 617-482-2700
F: 617-422-2910
URL: http://www.teradyne.com/

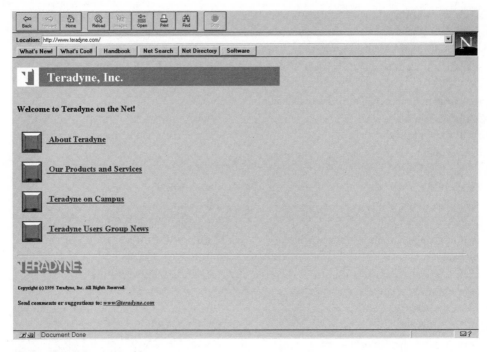

Figure 3-77 Teradyne.

TeraTech (*SW Utilities, Development*)
100 Park Ave., Suite 360
Rockville, MD 20850-2618
T: 301-424-3903
F: 301-762-8185
B: 301-762-8184
E-mail: info@teratech.com
FTP: ftp.teratech.com/pub/teratech/
URL: http://www.teratech.com/teratech/

Texas Industrial Peripherals
(*Peripherals, Input Devices*)
PO Box 49182
Austin, TX 78765
T: 512-837-0283
F: 512-837-0207
URL: http://www.ikey.com/

Texas Instruments (*ICs, Logic*)
Box 14149
12501 Research Blvd.
Austin, TX 78714-9149
T: 512-250-7111
FTP: ftp://ti.com/
URL: http://www.ti.com/

TexasMicro (*Computers, Systems*)
PO Box 42963
Houston, TX 77242-9910
T: 713-541-8200
F: 713-541-8226
E-mail: sales@texmicro.com

Texscan Corporation (*Equipment, Cable*)
4849 North Mesa St., Suite 200
PO Box 981006
El Paso, TX 79998

T: 800-351-2345
F: 915-591-6984
URL: http://www.texscan.com/

The Computer Journal   (*Publisher,*
Computers)
PO Box 3900
Citrus Heights, CA 95611-3900
T: 916-722-4970
F: 916-722-7480
B: 916-722-5799
E-mail: tcj@psyber.com
URL: http://www.psyber.com/~tcj/
URL: http://www2.psyber.com/~tcj/

Thinking Machines Corporation
(*Networking, Hardware and Software*)
14 Crosby Dr.
Bedford, MA 01730
T: 617-276-0400
F: 617-276-0444
URL: http://www.think.com/

Thomas & Betts   (*Manufacturers,*
Components)
(Refer to distributors for sales and techni-
cal information)
URL: http://www.sei-europe.com:8008/
sei/ans/ansmenu.htm

Time Electronics   (*Distributor,*
Components)
70 Marcus Blvd.
Hauppauge, NY 11788
T: 516-273-0100
F: 516-273-0136
URL: http://www.time.avnet.com/
(Includes links to nation-wide sales offices)

TLG Electronics   (*Manufacturer,*
Custom (*cable harness*))
1280 Alum Creek Dr.
Columbus, OH 43209
T: 614-252-1672
F: 614-252-6990
URL: http://www.infinet.com/~tlg/

TOA Electronics, Inc.   (*Peripherals,*
Speakers)
601 Gateway Blvd.
South San Francisco, CA 94080
T: 415-588-2538
F: 415-588-3349
E-mail: jpvelo@aol.com
URL:
http://www.sysint.com/systems/toa/

Tone Commander   (*Equipment,*
Communication)
11609 49th Place West
Mukilteo, WA 98275-4255
T: 206-349-1000
F: 206-349-1010
E-mail: tcs@halcyon.com
URL: http://www.halcyon.com/tcs/

Topdown   (*Simulation, IC Models*)
71 Spit Brook Rd., Suite 301
Nashua, NH 03060
T: 603-888-8811
F: 603-888-7694
URL: http://www.topdown.com/

Toray   (*Computers, Drives*)
1875 South Grant St., Suite 720
San Mateo, CA 94402
T: 415-341-7152
F: 415-341-0845
E-mail: info@toray.com
URL: http://www.toray.com/

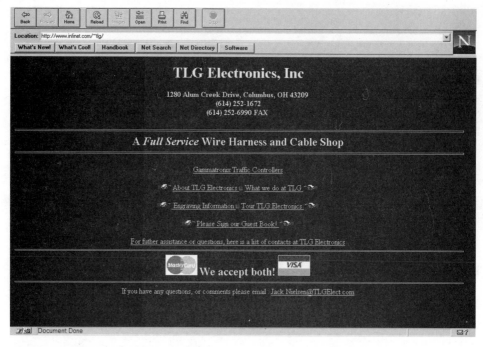

Figure 3-78 TLG Electronics.

Torch USA (*SW Utilities, Reference*)
PO Box 518
North Andover, MA 01845
T: 800-300-2199
F: 508-688-2199
URL: http://www.torch-usa.com/
URL: http://www.soshelpcenter.com/

Toshiba America (*Computers, Systems*)
9740 Irvine Blvd.
Irvine, CA 92718
T: 714-583-3000
F: 800-950-4373
B: 714-837-2116
URL: http://www.tais.com/
URL: http://www.toshiba.com/
(Includes links to worldwide business units)

Total Recall (*Services, Data Recovery/Diagnostics, Software*)
2462 Waynoka Rd.
Colorado Springs, CO 80915
T: 719-380-1616
F: 719-380-7022
URL: http://www.recallusa.com/

TouchStone Software Corporation
(*Diagnostics, Virus*)
Trend Micro Devices, Inc.
2130 Main St., Suite 250
Huntington Beach, CA 92648
T: 714-969-7746
F: 714-960-1886
URL: http://www.antivirus.com/

TransSwitch (*ICs, Communication*)
8 Progress Dr.
Shelton, CT 06484

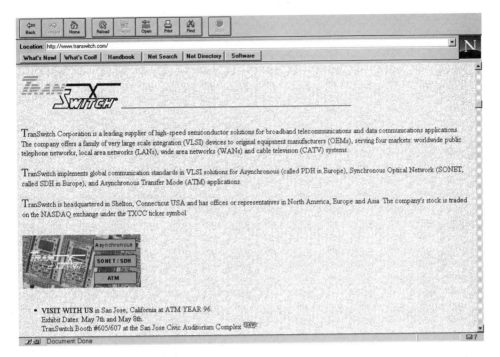

Figure 3-79 TransSwitch.

T: 203-929-8810
F: 203-926-9453
E-mail: appsengr@txc.com
URL: http://www.transwitch.com/

Traveling Software (*SW Utilities,*
Communication)
18702 North Creek Parkway
Bothell, WA 98011
T: 206-483-8088
F: 206-485-6786
B: 206-485-1736
URL: http://www.halcyon.com/
travsoft/
URL: http://www.travsoft.com/

Tredex (*Computers, Referbished*)
5306 Beethoven Ave.
Los Angeles, CA 90066

T: 310-301-0300
F: 310-301-0313
URL: http://www.tredex.com/

Triangle Electronics (*Distributor,*
Components)
T: 800-454-7722
E-mail: triangle@imsworld.com
URL: http://www.imsworld.com/
triangle/

Trilithic (*Manufacturer, Components*)
9202 East 33rd St.
Indianapolis, IN 46236
T: 317-895-3600
F: 317-895-3613
E-Mail: sales@trilithic.com
URL: http://www.trilithic.com/

TRW (*Manufacturer, Components*)
1900 Richmond Rd.
Cleveland, OH 44124
URL: http://www.trw.com/

TTi Technologies, Inc. (*ICs, BIOS*)
1445 Donlon St., #9
Ventura, CA 93003
T: 800-541-1943
E-mail: mike@hyperplex.com
URL: http://www.ttitech.com/

Tseng Labs, Inc. (*Controllers, Video*)
6 Terry Dr.
Newtown, PA 18940
T: 215-968-0502
F: 215-860-7713
B: 215-579-7536

TSI (*Equipment, Instruments*)
500 Cardigan Rd.
PO Box 64394
St. Paul, MN 55164
T: 612-483-0900
F: 612-490-2748
E-mail: info@tsi.com
URL: http://www.tsii.com/
URL: http://www.waldorf-gmbh.de/tsi/
homepage.html

Tucson Computer Corporation
(*Distributor, SIMM Adapters*)
6043 - 103 St.
Edmonton, AB Canada
T: 403-437-9797
F: 403-435-9400
E-mail: tucson@oanet.com
URL: http://www.tgx.com/tucson/

TurboSim (*Simulation, Circuit CAD*)
Island Logix, Inc.
PO Box 157
Waukegan, IL 60079

T: 708-360-0458
F: 708-360-0468
URL: http://pages.prodigy.com/L/J/A/
LJSN87A/

Turtle Beach Systems (*Peripherals,
Sound Boards*)
52 Grumbacher Rd.
York, PA 17402
T: 717-767-0200
F: 717-767-6033
B: 717-767-0250

Tyan Computer (*Distributor,
PC Assemblies*)
1645 South Main St.
Milpitas, CA 95035
T: 408-956-8000
F: 408-956-8044
B: 408-956-8171
FTP: ftp://ftp.tyan.com
Newsgroup:
alt.comp.periphs.mainboard.tyan
URL: http://www.tyan.com/

TyLink Corporation (*Services,
Education*)
10 Commerce Way
Norton, MA 02766
T: 508-285-0033
URL: http://204.178.65.252/

UB Networks (*Networking, Hardware*)
3900 Freedom Circle
Santa Clara, CA 95052-8030
T: 408-496-0111
F: 408-970-7343
URL: http://www.ub.com/

Ultrastor Corporation (*Controllers,
Drive and SCSI*)
15 Hammond St., Suite 310
Irvine, CA 92718

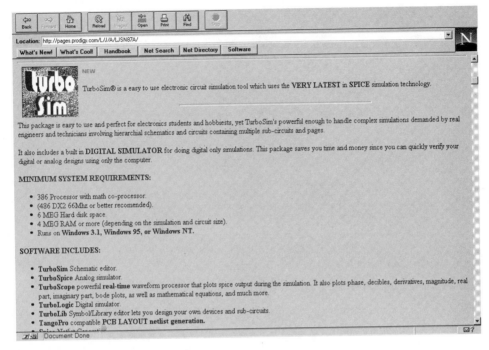

Figure 3-80   TurboSim.

T: 714-581-4100
F: 714-581-0826
B: 714-581-4125
E-mail: ustor@primenet.com
FTP:
ftp://ftp.primenet.com/users/u/ustor

UMAX Technologies   (*Computers, Systems*)
T: 800-562-0311
F: 310-523-9399
FaxBack: 510-651-3710
B: 510-651-2550

Unicon, Inc.   (*Equipment, Communication*)
4246 Park Glen Rd.
Minneapolis, MN 55416
T: 800-747-9615

URL: http://www.telephone-headset.com/index.html

Unicore Software Inc.   (*ICs, BIOS*)
1538 Turnpike St.
N. Andover, MA 01845
T: 508-686-6468
F: 508-683-1630
URL: http://www.unicore.com/

Unisys Corporation   (*Computers, Systems*)
PO Box 500
Blue Bell, PA 19424
T: 215-986-4011
URL: http://www.unisys.com/

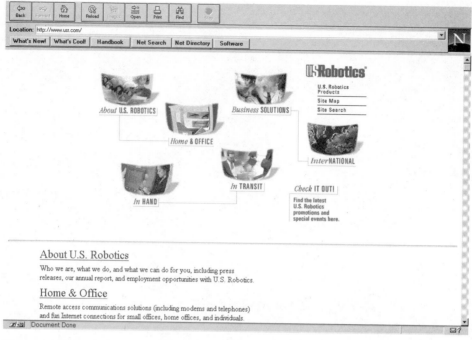

Figure 3-81 U.S. Robotics.

Upgrades Unlimited (*Distributor, Memory*)
PO Box 815034
Dallas, TX 75381
T: 214-243-5808
F: 214-243-6823
E-mail: uui@globallink.net
URL: http://www.globallink.net/~uui/

US Computer Direct Corporation
(*Distributor, PC Assemblies*)
16291 Gothard St.
Huntington Beach, CA 92647
T: 714-841-4081
F: 714-841-4262
E-mail: uscomputer@prtcl.com
URL: http://uscomputer.prtcl.com/

U.S. Robotics (USR) (*Computers, Modems*)
8100 N. McCormick Blvd.
Skokie, IL 60076
T: 708-982-5151
FaxBack: 800-762-6163
B: 708-982-5092
CIS: go USROBOTICS
E-mail: support@usr.com
URL: http://www.usr.com/

U.S. Software (*SW Utilities, Embedded Product Development*)
14215 NW Science Park Dr.
Portland, OR 97229
T: 503-641-8446
F: 503-644-2413
E-mail: support@ussw.com
URL: http://www.ussw.com/

U.S. Technologies, Inc. (*Manufacturer, PCB*)
P.O. Box 697
Lockhart, TX 78644
T: 512-376-1049
F: 512-376-1042
E-mail: ustech@usxx.com
URL: http://www.usxx.com/

UUNet Technologies (*Services, On-line*)
3060 Williams Dr.
Fairfax, VA 22031-4648
T: 703-206-5600
F: 703-206-5601
E-mail: info@uu.net
URL: http://www.uu.net/index.html

V3 Semiconductor (*ICs, Logic*)
T: 408-988-1050
F: 408-988-2601
E-mail: v3help@vcubed.com
URL: http://www.vcubed.com/
(Includes links to worldwide distributors)

Varian (*Equipment, Instruments*)
3120 Hansen Way, D-103
Palo Alto, CA 94304-1030
T: 415-424-5710
F: 415-855-7201
URL: http://www.varian.com/

Vector Electronics Company
(*Equipment, Prototyping*)
12460 Gladstone Ave.
Sylmar, CA 91342
T: 818-365-9661
F: 818-365-5718
E-mail: inquire@vectorelect.com
URL: http://www.vectorelect.com/

Vektron International (*Computers, Systems*)
2100 N. Hwy. 360, Ste 1904
Grand Prairie, TX 75050
T: 214-606-0280
F: 214-606-1278
CIS: go VEKTRON
URL: http://www.vektron.com/

Ventana Communications Group
(*Publisher, Computer*)
P.O. Box 13964
Research Triangle Park, NC 27713-3964
T: 919-544-9404
E-mail: feedback@vmedia.com
URL: http://www.vmedia.com/

VenturCom, Inc. (*SW Utilities, Development*)
215 First St.
Cambridge, MA 02142
T: 617-661-1230
F: 617-577-1607
E-mail: info@vci.com
FTP: ftp://ftp.vci.com/
URL: http://www.vci.com/

Verbex Voice Systems (*SW Utilities, Voice Recognition*)
1090 King Georges Post Rd.
Edison, NJ 08837-3701
T: 908-225-5225
F: 908-225-7764
E-mail: sales@listen.verbex.com
URL: http://www.txdirect.net/verbex/

VeriBest, Inc. (*Simulation, Circuit CAD*)
6101 Lookout Rd.
Boulder, CO 80301
T: 800-837-4237
URL: http://www.veribest.com/

Figure 3-82   Ventana Communications Group.

Verilog, Inc.   (*SW Utilities, Embedded Product Development*)
3010 LBJ Freeway, Suite 900
Dallas, TX 75234
T: 214-241-6595
F: 214-241-6594
E-mail: info@logtech.com
URL: http://www.verilogusa.com/

Vicor Corporation   (*Equipment, Power Products*)
23 Frontage Rd.
Andover, MA 01810
T: 508-470-2900
F: 508-475-6715
URL: http://www.vicr.com/

VideoLabs   (*Equipment, Video*)
10925 Bren Rd. E.
Minneapolis, MN 55343

T: 612-988-0055
F: 612-988-0066
E-mail: videolabs@flexcam.com
URL: http://www.flexcam.com/

Videonics   (*Equipment, Video*)
1370 Dell Ave.
Campbell, CA 95008
T: 408-866-8300
F: 408-866-4859
URL: http://www.videonics.com/

VIEWlogic   (*Simulation, Circuit CAD*)
293 Boston Post Rd. West
Marlborough, MA 01752
T: 508-480-0881
F: 508-480-0882
URL: http://www.viewlogic.com/

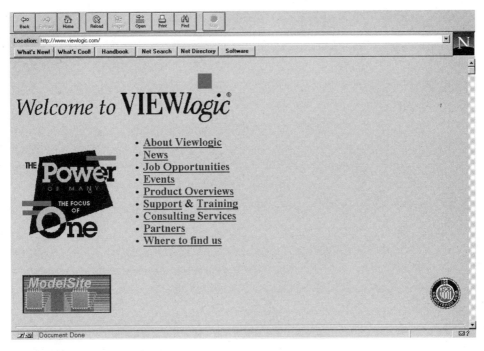

Figure 3-83 VIEWlogic.

ViewSonic Corporation (*Peripherals, Monitors*)
20480 Business Pkwy.
Walnut, CA 91789
T: 909-869-7976
B: 909-468-1241
CIS: 73374,514
URL: http://www.viewsonic.com/

Virtual Machine Works (*Equipment, Emulation*)
One Kendall Square, Building 600
Cambridge, MA 02139
T: 617-621-1700
F: 617-621-7927
URL: http://www.vmw.com/

Visio Corporation (*SW Utilities, Business*)
520 Pike St., Suite 1800
Seattle, WA 98101
T: 206-521-4600
F: 206-521-4601
FaxBack: 206-521-4550
CIS: 74774,2161
URL: http://www.visio.com/

Visiontek (*Distributor, Memory*)
URL: http://www.visiontek.com/

Visionware Ltd. (*SW Utilities, Development*)
Conqueror House
Vision Park, Cambridge, CB4 4ZR UK
T: +44 (0) 1223 518000
F: +44 (0)1223 518001
URL: http://www.visionware.com/

Visioneer (*Distributor, Scanners*)
2860 West Bayshore Rd.
Palo Alto, CA 94303
T: 415-812-6400
FaxBack: 800-505-0175
E-mail: 4info@visioneer.com
URL: http://www.visioneer.com/

Visionics (*Simulation, Circuit CAD*)
995 E. Baseline Rd., Ste. 2166
Tempe, AZ 85283-1336
T: 602-730-8900
F: 602-730-8927
E-mail: visioneda@aol.com
URL: http://www.bahnhof.se/~visionics/

Visual Communications Company
(*Manufacturer, Components*)
T: 800-777-7334
URL: http://www.vcclite.com/
(Includes links to nation-wide distributors)

Vive Synergies, Inc. (*Equipment, Communication*)
30 West Beaver Creek Rd., Unit 101
Richmond Hill, Ontario, L4B 3K1
Canada
T: 905-882-6107
F: 905-882-6238
URL: http://www.vive.com/

Voxel (*Equipment, Imaging*)
26081 Merit Circle #117
Laguna Hills, CA 92653
T: 714-348-3200
F: 714-348-8665
E-mail: info@voxel.com
URL: http://www.voxel.com/

Wall Data, Inc. (*Networking, Hardware and Software*)
11332 NE 122nd Way
Kirkland, WA 98034-6931

T: 206-814-9255
URL: http://www.walldata.com/

Walnut Creek CD-ROM (*Publisher, Software*)
URL: http://www.cdrom.com/

Watcom International (*SW Utilities, Development*)
Powersoft Division
561 Virginia Rd.
Concord, MA 01742
T: 508-287-1500
URL: http://www.watcom.com/

WaterGate Software (*Diagnostics, Software*)
2000 Powell St., Suite 1200
Emeryville, CA 94608
T: 510-596-1770
F: 510-653-4784
E-mail: support@ws.com
URL: http://www.ws.com/

Wave Tech (*Diagnostics, Software*)
9145 Balboa Ave.
San Diego, CA 92123
T: 619-279-2200
F: 619-627-0132
B: 619-278-5034

Webb Distribution (*Distributor, Components*)
2 Lowell Ave.
Winchester, MA 01890
T: 617-729-5800
F: 617-729-6839
E-mail to: sales@webbdistinc.com
URL: http://www.webbdistinc.com/

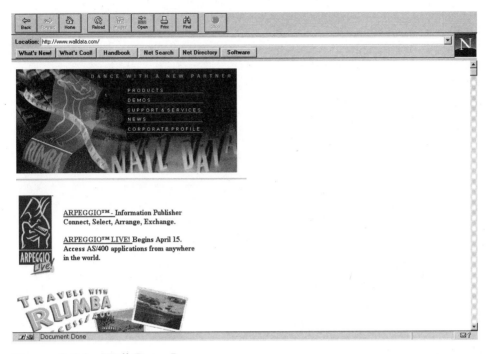

Figure 3-84 Wall Data, Inc.

Web Laboratories (*Simulation, Circuit CAD*)
13731 W. Capitol Dr., Suite 260
Brookfield, WI 53005
T: 414-367-6825
F: 414-367-6824
E-mail: info@webblabs.com
URL: http://www.webblabs.com/

Weitek Corporation (*ICs, Microprocessors*)
1060 E. Arques
Sunnyvale, CA 94086
T: 408-526-0300
F: 408-738-1185
B: 408-522-7512
E-mail: support@weitek.com
URL: http://www.weitek.com/

Wen Technology (*Peripherals, Monitors*)
T: 1-800-377-4-WEN
F: 914-347-4128
E-mail: Wentech@internetmci.com
URL: http://www.iiactive.com/wen/default.htm

Wenzel Associates (*Manufacturer, Components*)
1005 La Posada Dr.
Austin, TX 78752
T: 512-450-1400
F: 512-450-1490
URL: http://www.wenzel.com/

Figure 3-85 Winbond Electronics Corporation.

Western Digital Corporation
(*Computers, Drives*)
8105 Irvine Center Dr.
Irvine, CA 92718
T: 714-932-5000
F: 714-932-4012
B: 714-753-1068
CIS: go IBMHW
FTP: ftp.wdc.com
URL: http://www.wdc.com/
URL: http://www.wdc.com:80/support/

Western Micro Technology
(*Networking, Hardware and Software*)
254 E. Hacienda Ave.
Campbell, CA 95008
T: 408-379-0177
F: 408-341-4762
URL: http://www.westernmicro.com/

Winbond Electronics Corporation
(*ICs, Communications*)
No.4, Creation Rd. 3
Science-Based Industrial Park
Hsinchu, Taiwan
T: 886-35-770066
F: 886-35-792668
URL: http://www.winbond.com.tw/

Windows Magazine   (*Publisher, Computer*)
URL: http://www.winmag.com/

Windows Memory Corporation
(*Distributor, Memory*)
920 Kline, Suite 302
La Jolla, CA 92037
T: 619-454-9701
F: 619-454-9703
E-mail: winmem@crash.cts.com

Wind River Systems (*SW Utilities, Development*)
1010 Atlantic Ave.
Alameda, CA 94501
T: 510-748-4100
F: 510-814-2010
E-mail: inquiries@wrs.com
URL: http://www.wrs.com/

WinZip (*SW Utilities, Compression*)
Nico Mak Computing, Inc.
PO Box 919
Bristol, CT 06011
E-mail: sales@compuserve.com
URL: http://www.winzip.com/

WordPerfect (*SW Utilities, Business*)
1555 N. Technology Way
Orem, UT 84057
T: 801-225-5000
F: 801-229-1566
B: 801-225-4414
URL: http://www.wordperfect.com/

WSI, Inc. (*ICs, Logic*)
T: 510-656-5400
F: 510-657-8495
URL: http://www.wsipsd.com/

Wyse Technology (*Peripherals, Monitors*)
3471 N. 1st St.
San Jose, CA 95134
T: 408-473-1200
F: 408-473-1972
B: 408-922-4400
URL: http://www.wyse.com/wyse/

Xerox (*Computers, Systems*)
Xerox Square
Rochester, NY 14644
T: 716-423-5078

F: 203-968-3368
URL: http://www.xerox.com/

Xicor (*ICs, Logic*)
1511 Buckeye Dr.
Milpitas, CA 95035
T: 408-432-8888
F: 408-432-0640
B: 408-943-0655
FaxBack: 408-954-1627
E-mail: info@smtpgate.xicor.com
URL: http://www.xicor.com/

Xilinx (*ICs, Logic*)
2100 Logic Dr.
San Jose, CA 95124
T: 408-559-7778
F: 408-879-4676
URL: http://www.xilinx.com/

Xionics (*Equipment, Imaging*)
70 Blanchard Rd.
128 Corporate Center
Burlington, MA 01803
T: 617-229-7000
F: 617-229-7119
E-mail: info@xionics.com
URL: http://www.xionics.com/

Xircom (*Networking, Hardware*)
26025 Mureau Rd.
Calbasas, CA 91302
T: 805-376-9200
F: 805-376-9220
B: 818-878-7618
URL: http://www.organic.comAds/
Xircomindex.html
URL: http://www.organic.com/
Commercial/Xircom/

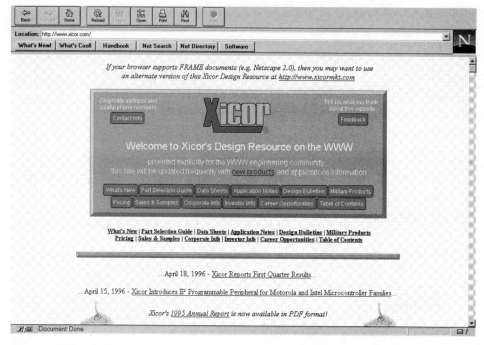

Figure 3-86 Xicor.

Xylogics, Inc. (*Networking, Hardware*)
53 Third Ave.
Burlington, MA 01803
T: 617-272-8140
F: 617-273-5392
URL: http://www.xylogics.com/

Zenith Data Systems (*Computers,
Systems*)
2150 E. Lake Cook Rd.
Buffalo Grove, IL 60089
T: 800-553-0331
F: 800-472-7211
B: 708-808-2264
URL: http://www.zds.com/

Zenon Technology, Inc. (*Computers,
Systems*)
T: 800-899-6119
E-mail: sales@zenontech.com

URL:
http://www.primenet.com/~zenon/

Zeos International, Ltd. (*Computers,
Systems*)
1301 Industrial Blvd.
Minneapolis, MN 55413
T: 612-362-1234
F: 612-362-1207
B: 612-362-1219
CIS: go ZEOS
E-mail: support@zeos.com

Ziatech Corporation (*Computers,
Industrial*)
1050 Southwood Dr.
San Luis Obispo, CA 93401
T: 805-541-0488
F: 805-541-5088
E-mail: info@ziatech.com
URL: http://www.ziatech.com/

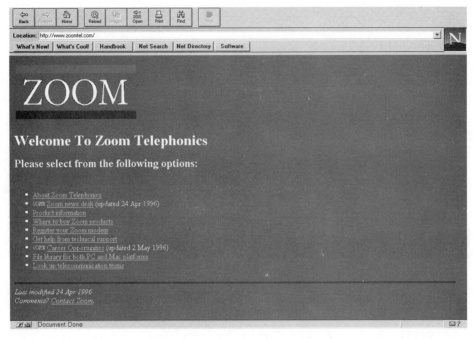

Figure 3-87   Zoom Telephonics.

Zilog   (*ICs, Logic*)
210 East Hacienda Ave.
Campbell, CA 95008
T: 408-370-8246
F: 408-370-8056
URL: http://www.zilog.com/

Zoom Telephonics   (*Computers, Modems*)
207 South St.
Boston, MA 02111
T: 617-423-1076
F: 617-423-5536
B: 617-423-3733
FaxBack: 617-423-4651
AOL: keyword ZOOMT
CIS: go ZOOM
URL: http://www.zoomtel.com/

Z-World Engineering   (*Equipment, Industrial*)
1724 Picasso Ave.
Davis, CA 95616-0547

T: 916-757-3737
F: 916-753-5141
E-mail: zworld@zworld.com
URL: http://www.zworld.com/zworld/

Zylab   (*SW Utilities, Imaging*)
19650 Club House Rd., Suite 106
Gaithersburg, MD 20879
T: 301-590-0900
URL: http://www.zylab.com/

ZyXEL   (*Peripherals, Modems*)
4920 E. La Palma Ave.
Anaheim, CA 92807
T: 714-693-0808
F: 714-693-8811
B: 714-693-0762
URL: http://www.zyxel.com/

GENERAL/REFERENCE CONTACTS (ALPHABETICAL BY GENERAL TOPIC)

8051 FAQ (and other PC FAQs)

ftp.rtfm.mit.edu/pub/usenet/
sci.answers/microcontroller-faq/8051
http://archive.cis.ohio-state.edu/
hypertext/faq/bngusenet/comp/
lang/forth/top.html
http://www.ece.orst.edu/serv/
8051/
8051 Home Page

Animation

See Graphics.

ARJ Compression Support and Information

See Compression Information.

ASCII Art FAQ

http://gagme.wwa.com/~boba/
faq.html

Audio File Format FAQ

http://www.cis.ohio-state.edu/
hypertext/faq/usenet/audio-fmts/
top.html

Benchmarks FAQ

http://hpwww.epfl.ch/bench/bench.
FAQ.html

Careers

See Jobs and Careers.

CD-ROMs

See Networked CD-ROM FAQ.

Circuit Simulation

See Simulation.

Classic Computers

http://deja-vu.oldiron.cornell.edu/

Clock Speed Adjustments on a Mac

http://violet.berkeley.edu/~schrier/
mhz.html

CMOS and System Optimization Information

http://www.dfw.net/%7Esdw/

Code and Cryptography FAQ

http://archive.cis.ohio-state.edu/
hypertext/faq/bngusenet/sci/crypt/
top.html

Colleges

See Universities.

Communication

http://www.analysis.co.uk/
commslib.htm
Communication & Telecommunication Library

http://www.cmpcmm.com/cc/
standards.html
Computer and Communication Standards

Components

http://www.bdi.ie/bdi/directs/2.htm
Business Directory International –
Electronic Components

http://www.entrepreneurs.net/
wondarl/eps.htm
Electronic Parts Supply

http://204.214.110.250/part-net/
welcome.htm
Part Net International

Compression Information

http://www.Dunkel.de/ARJ
Welcome to ARJ

Computer Acronym/Lexicon/ Symbol Index

http://rel.semi.harris.com/docs/
lexicon/
Lexicon of Semiconductor Terms

http://www.hike.te.chiba-u.ac.jp/
iec417/index.html
IEC Standard 417 Electronic Symbol
Index

http://www.zdnet.com/~pcmag/issues/
1501/pcm00120.htm
Everyone's Guide to Computer
Acronyms (A - N)

http://www.zdnet.com/~pcmag/issues/
1502/pcm00080.htm
Everyone's Guide to Computer
Acronyms (N - Z)

Computer Events

http://www.kweb.com/contents.html

Computer Graphics

See Graphics.

Computer Science Information and Links

http://src.doc.ic.ac.uk/bySubject/
Computing/Overview.html
WWW Computing Library

http://www.utexas.edu/computer/vcl/
Virtual Computer Library

Connecting MIDI Devices

http://harmony-central.mit.edu/
MIDI/
See Music Circuits.

Controls (Control Engineering)

http://www-control.eng.cam.ac.uk/
extras/Virtual_Library/
Control_VL.html

CPUs

See Microprocessor Information.

Datasheets

See IC Directory On-line and Other
IC Search Engines.
See Components.

Dictionaries

http://c.gp.cs.cmu.edu:5103/prog/
webster/
Webster's Dictionary

http://wombat.doc.ic.ac.uk/
Computing Dictionary

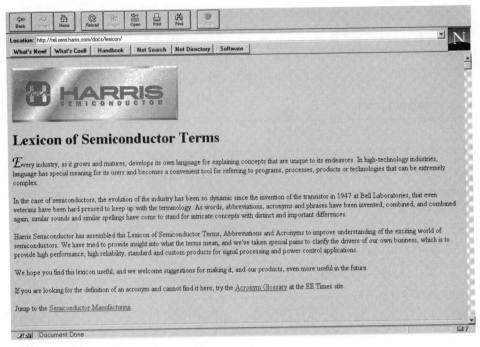

Figure 3-88 Sample on-line lexicon.

http://www.ll.mit.edu/ComLinks/
deskref.html#DTA
Dictionaries, Thesauri, & Acronyms

See On-line Dictionary of Computing.

Digital Signal Processing (DSP)

ftp://ftp.ti.com/mirrors/tms320bbs/

http://spib.rice.edu:80/spib.html
Signal Processing Database

http://tjev.tel.etf.hr/DSP/sigproc/

http://www.bdti.com/
Berkeley Design Technology

http://www.cera2.com/dsp.htm
High-Performance DSP Systems

http://www.dspnet.com/
DSP Net

EIA (Electronic Industries Association) - Consumer Electronic Manufacturer's Association

See Professional Organizations.

Electrical Engineer's Circuit Cookbook/Archives and Link Lists

ftp://ftp.armory.com/pub/user/
rstevew/

ftp://ftp.best.com/pub/cera/

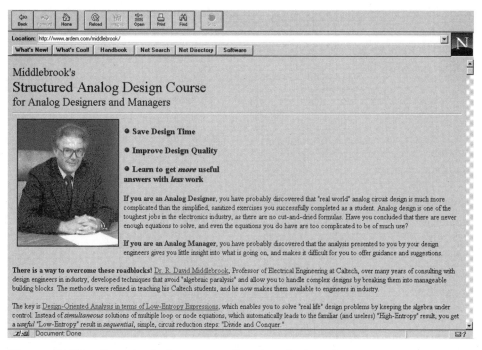

Figure 3-89 Sample design course for electrical engineers.

ftp://ieee.cas.uc.edu/pub/electronics/
mirrors/

http://arioch.gsfc.nasa.gov/paws/html/
homepage.html
NASA Parts Analysis Web System

http://cal003109.student.utwente.nl/
stefan/el.html
We-Man's Electro Stuff

http://engr-www.unl.edu/ee/eeshop/
netsites.html
E.E. Internet Info Sites

http://epims1.gsfc.nasa.gov/
engineering/ee.html
Electrical Engineering Library

http://epims1.gsfc.nasa.gov/
engineering/engineering.html
General Engineering Library

http://microship.ucsd.edu/
Nomadic Research Labs

http://sleepless.cs.uiuc.edu/sigarch/

http://weber.u.washington.edu/~pfloyd/
ee/index.html
Electrical Engineering Circuits Archive

http://www.ardem.com/middlebrook/
*Middlebrook's Structured Analog
Design Course*

http://www.avtechpulse.com/faq.html
The Unusual Diode FAQ

http://www.cadvision.com/guide/
science/electron/electron.html
Electrical Engineering Links
[exceptional resource]

http://www.cera2.com/ebox.htm
Electronic Engineers' Toolbox

http://www.circuitworld.com/
CircuitWorld Online Services

http://www.ecn.uoknor.edu/~jspatric/
ee-info.html
John's New Electrical Engineering Page

http://www.ee.ualberta.ca/html/
cookbook.html
Electronic Cookbook Archive

http://www.ee.umr.edu/corp/
Miscellaneous EE Links

http://www.ee.umr.edu/schools/
ee_programs.html
Index of Worldwide EE Programs

http://www.enc.hull.ac.uk/AP/G26/
The II-VI Semiconductors Home Page

http://www.engineers.com/tec.html
The Engineer's Club

http://www.industry.net/
IndustryNet

http://www-mtl.mit.edu/
semisubway.html
The Semiconductor Subway

http://www.smartlink.net/~bmcd/
semi/cat.html
*SPECNet: Semiconductor Process &
Equipment Network*

http://www.ucsalf.ac.uk/~bens/
eleclink.htm
Electronic and Related Links

http://www.washington.edu:1180/
tech_home/
*Computing and Information
Technologies*

Electronic Parts and Materials

http://whirligig.ecs.soton.ac.uk/
~ajb92/electron/data.html
Data Sheets Online

http://www.crhc.uiuc.edu/~dburke/
databookshelf.html
Data Bookshelf

http://www.paranoia.com/~filipg/
HTML/LINK/F_mail_order.html
*List of Mail Order Electronics
Companies*

http://www.part.net:80/
PartNet

Employment

See Jobs and Careers.

Federal Communications Commission (FCC)

gopher://gopher.fcc.gov/

General Electronic Information and Sites

ftp://ftp.netcom.com/pub/di/
dibald/
Schematic Directory

Figure 3-90 Sample link list.

ftp://wuarchive.wustl.edu/pub/
science/electrical/

http://ev-www.ser.fm.uit.no/electronics/
index.html
*List of Information Related to Elec-
tronic Engineering*

http://iquest.com/~virtual/max/
bebop.html
Boolean Logic Tutorial

http://iquest.com/~virtual/max/
mainmax.html
Virtual Hardware Site

http://www.commerce.net/
Commerce Net

http://www.e2w3.com/

http://www.ping.be/~ping0751/
home.htm
The Electronics HomeWorld

http://www-
soe.stanford.edu/soe/ieee/eesites.html
EE/CS Mother Site

http://www.std32.com/
STD32 Site

http://www.teleport.com/~usb/
Universal Serial Bus Site

General PC Information

http://engr-www.unl.edu/ee/
eeshop/miscinfo.html
Miscellaneous Project Information

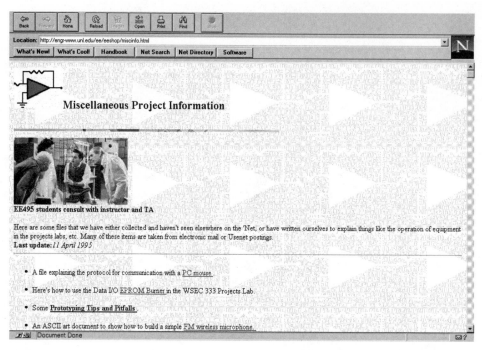

Figure 3-91 General PC information.

http://gnn.com/wic/wics/comput.new.
html
GNN Select Computer Page

http://www.compinfo.co.uk/
tphdwr.htm#T-STORAGE
*CIC Hardware Topics [exceptional
resource]*

http://web2.airmail.net/~admin/
index.htm
Expert Solutions

http://www.guernsey.net/~pad/
index.html
Electronic Repair Tips Home Page

General Software

See Utility Software.

Graphics

http://dragon.jpl.nasa.gov/~adam/
transparent.html
*The Transparent/Interlaced GIF
Resource Page*

http://www.cis.ohio-
state.edu/hypertext/faq/usenet/graphics
/animation-faq/faq.html
comp.graphics.animation FAQ

http://www.infomedia.net/scan
The Scanning FAQ

Hard Drive Information

http://theref.c3d.rl.af.mil/
TheRef: Drive and Controller Guide

http://www.cs.yorku.ca/People/frank/
docs/docs-hd.html
Hard Drive Documents

http://www.wi.leidenuniv.nl/ata/
*The comp.sys.ibm.pc.hardware.storage
WWW Site*

Hardware/Software Vendor List

http://www.syssvc.com/vendor.html

Hobby Information

See Science Hobby Sites.

Home Automation

http://www.hometeam.com/
The Home Team

http://www.lonworks.echelon.com
Echelon Home Automation

HTML File with Over 300 PC-Related Sites

Send an e-mail message to:
      lejeune@acy.digest.net
In the body of the message, type:
      send urb-menu.htm

IBM Keyboard FAQ (and Other FAQs)

ftp.armory.com/pub/user/rstevew/

IC Data

See IC Directory On-line and Other IC Search Engines.

IC Directory On-line and Other IC Search Engines

http://einstein.et.tudelft.nl/
~offerman/chiplist.html
CHIPLIST 9.2 [exceptional resource]

http://hearstelectroweb.com/
EEM and IC Master

http://www.hitex.com:80/chipdir/
chipdir.html
Chip Directory

http://www.paranoia.com/~filipg/
HTML/cgi-bin/giicm_form.html
IC Search Site

http://www.questlink.com/
Questlink Technology

IC Fabrication Tutorial

http://rel.semi.harris.com/docs/
lexicon/manufacture.html
"How Semiconductors Are Made"

IDE Information

http://curia.ucc.ie/cgi-bin/
acronym/ide
Acronyms and Abbreviations

http://http.tamu.edu:8000/~jdb8042/
SmallSys/8bitIDE.html
*Connecting IDE Devices to 8-bit
Machines*

http://www.paranoia.com:80/~filipg/
HTML/LINK/F_IDE-tech.html
IDE - Hardware Reference & Information Document Ver 1.00

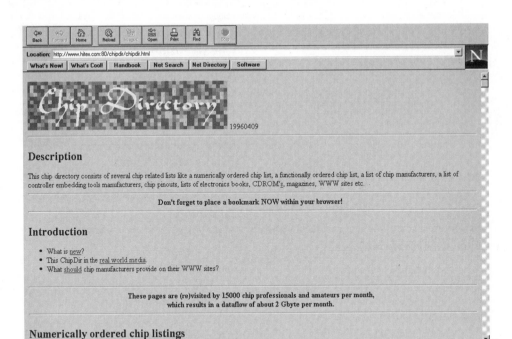

Figure 3-92   Chip directory.

http://www.wi.leidenuniv.nl/ata/
atafq.html
*The Enhanced IDE/Fast-ATA/ATA-2
FAQ*

See Hard Drive Information.

IEEE - Institute of Electrical and Electronics Engineers

See Professional Organizations.

Internet Resources

ftp://geec-gw.njit.edu/pub/dns/
cptd.faq
*Explanation of the Internet Domain
Name System (DNS)*

http://archive.cis.ohio-state.edu/
hypertext/information/rfc.html
Internet Request For Comments (RFC)

http://chico.ncsa.uiuc.edu/SDG/
Software/WinMosaic/FAQ.html
The Moasia FAQ

http://hoohoo.ncsa.uinc.edu/ftp/
FTP interface FAQ

http://ici.proper.com/
HTML Publishing Info.

http://www.boutell.com/faq/
World Wide Web FAQ

http://theory.cd.uni-bonn.de/
ppp/faq.html
Internet Point-to-Point [PPP] FAQ

http://www.amazing.com/internet/
The Internet Access FAQ - How to become a provider

http://www.brandonu.ca/~ennsnr/
Resources/Welcome.html
Internet Resources

http://www.cis.ohio-state.edu/
hypertext/faq/usenet/
FAQ-List.html
Usenet FAQs

http://www.cis.ohio-state.edu/
hypertext/faq/usenet/
internet-services/top.html
USENET FAQs: Internet Services

http://www.cis.ohio-state.edu/
hypertext/faq/usenet/mail/
filtering-faq/faq.html
Filtering Mail FAQ

http://www.eit.com:80/web/
netservices.html
Explaining Internet Resources

http://www.excite.com/Subject/
Computing/Access_Providers/
s-index.h.html
An index of national and international Internet service providers

http://www.geopages.com/SiliconValley/
1612/
Extensive Internet Links

http://www.reach.com/matrix/
Into the Matrix

http://www.w3.org/
The World Wide Web Consortium

ISO - International Standards Organization

See Professional Organizations.

Jobs and Careers

http://helpwanted.com/
Recruitment Online, Inc.

http://internet-plaza.net/careermag/
NCS Career Magazine

http://none.coolware.com/jobs.html
Coolware Electronic Job Guide

http://stimpy.cen.uiuc.edu/comm/
expo/
Engineering Employment Expo

http://www.careermag.com/careermag/
news/index.html
Career Magazine Jobline Database

http://www.careermosaic.com/cm/
Career Mosaic

http://www.chicago.tribune.com/
career/
Chicago Tribune Career Finder

http://www.monster.com/
The Monster High-Tech Job Board

Macintosh Information

ftp://ftp.ari.net/pub/MacSciTech/
elecEng/

Magazines

See News Contacts and Sites.

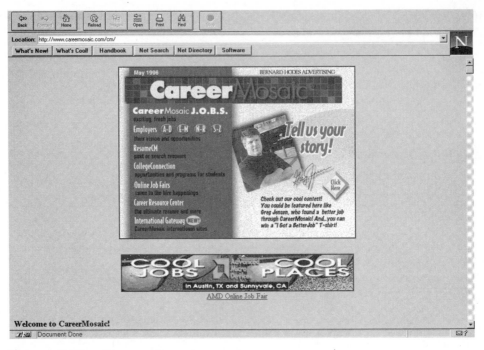

Figure 3-93 Career reference.

Metalworking Information: Soldering, Drilling, Welding, Cutting, etc.

http://www.paranoia.com:80/
~filipg/HTML/LINK/
Metal_idx.html
rec.crafts.metalworking FAQ

Microprocessor Information and Assembler Utilities

ftp://ftp.evans.ee.adfa.oz.au/pub/
micros/

ftp://ftp.funet.fi/pub/microprocs/

ftp://ftp.luth.se/pub/misc/microchip/

ftp://ftp.luth.se/pub/misc/microchip/
third-party/others/pic84faq.zip

ftp://ftp.luth.se/pub/misc/motorola/

ftp://ftp.pppl.gov/pub/8051/
signetics-bbs/

ftp://ftp.ultranet.com/biz/mchip/

ftp://evans.ee.adfa.oz.au/pub/micros/

ftp://rtfm.mit.edu/pub/usenet/
comp.answers/microcontroller-faq/
primer/

http://infopad.eecs.berkeley.edu/CIC/
CPU Information Center

http://livewire.ee.latrobe.edu.au/~sjm/8
051.html

http://me210.stanford.EDU/luehrb/
uP.html
Microprocessor Page

http://rasi.lr.ttu.ee/~sis/
Computer Page

http://www.cera2.com/gator.htm
US Software navi-GATOR

http://www.cera2.com/micro.htm
Microcontroller/Microprocessor Internet Resource List

http://www.ece.orst.edu/~paul/
8051-goodies/goodies-index.html
Paul's 8051 Microcontroller Family

http://www.ee.nmt.edu/ee/hc11/
homepage.html
Motorola 68HC11 Code Library

http://www.ee.ualberta.ca/archive/
m68kfaq.html
comp.sys.m68k FAQ

http://www.kaiwan.com/~douglas/
Black Feather Electronics Info Page

http://www.lancs.ac.uk/people/
cpaame/pic/pic.htm
Andy's PIC Project Page

http://www.panix.com/stimpson/
micro.html
The Versatile 8031 (or 8051) Microcontroller

Modems

http://elaine.teleport.com/~curt/
modems.html
Curt's High Speed Modem Page

http://www.cmpcmm.com/cc/
Computer and Communication relevant URLs

Monitor Setups and Troubleshooting Information

http://www.devo.com/video
Fixed Frequency PC Video FAQ

http://www.pc.ibm.com/monitors/
letter.html
IBM notifies customers of 9527 color monitor repair

Motors and Control Information

http://www.cs.uiowa.edu/~jones/
setp/
Stepping Motor Control

Music Circuits

http://rowlf.cc.wwu.edu:8080/
~n9343176/schems.html
Music Electronics Archive

http://users.aol.com/jorman/
The Digital Music Zone

http://www.paranoia.com:80/~filipg/
HTML/LINK/F_MIDI.html
MIDI - MUSICAL INSTRUMENT DIGITAL INTERFACE V1.00

Networked CD-ROM FAQ

http://saturn.uaamath.alaska.edu/
~gibbsg/cdromlan_faq.html
CDROMLAN FAQ

Networking Information

http://archive.cis.ohio-
state.edu/hypertext/faq/bngusenet/
comp/protocols/snmp/top.html
USENET FAQs: comp.protocols.snmp

http://garage.ecn.purdue.edu/~papers/
PAPERS Page

News Contacts and Reference Sites

http://192.216.191.71/ebn/
current/
Electronic Buyers News Online

http://nytsyn.com/cgi-bin/
times/lead/go/
New York Times: Computer News Daily

http://tcp.ca/
The Computer Paper

http://techweb.cmp.com/techweb/
techweb/current/
*Windows Mag., Home PC, NetGuide,
and more*

http://www.bookmasters.com/edt.htm
Electronic Distribution Today

http://www.cnn.com/TECH/index.html
CNN Technology News

http://www.ednmag.com/
EDN - Electronic Design News

http://www.emmonline.com/
EEM - Electronic Engineer's Master

http://www.elsevier.nl:80/eee/
Menu.html
*Electrical and Electronic Engineering
Alert*

http://www.hotwired.com/Login/
HotWired magazine

http://www.infi.net/naa/hot.html
Newspapers on the Web

http://www.mwjournal.com/
Microwave Journal, Journal of Electronic Defense, and more

http://www.pcinews.com/pci/
*Computer Trade Show Calendar &
Industry News*

http://www.senn.com/
Science & Engineering Network News

http://www.sumnet.com/enews
Electronic News

http://www.techtalkmag.com/
Appliance Tech Talk

http://www.unitedmedia.com/comics/
dilbert/
Dilbert Comic

http://www.upside.com/
Upside Magazine

http://www.vbonline.com/vb-mag/
Visual Basic Online

http://www.videomaker.com/
VideoMaker Online

http://www.yahoo.com/headlines/
compute/
Yahoo Tech Summary

http://www.zdnet.com/~pccomp/
PC Computing

http://www.zdnet.com/~pcmag/
PC Magazine

http://www.ziff.com/Welcome_
noimap.html
Ziff-Davis Publications

Figure 3-94   News digest.

http://www2.nando.net/nt/info/
The Nando Times

Newsgroup Creation FAQ

http://www.math.psu.edu/barr/
alt-creation-guide.html

Newsletters

See News Contacts and Sites.

Novell Labs Hardware Certification Database

http://www.novell.com/ServSupp/
labsform.html

On-line Dictionary of Computing

See Dictionaries.

Organizations

See Professional Organizations.

OSI (Open Systems Interconnection) FAQ

http://archive.cis.ohio-state.edu/
hypertext/faq/usenet/osi-protocols/
faq.html
comp.protocols.iso FAQ

Parallel Port Documents

ftp://ftp.ee.ualberta.ca/pub/
cookbook/comp/ibm/
pport094.doc.z

http://www.paranoia.com:80/~filipg/
HTML/LINK/PORTS/F_Parallel.html
PC Parallel/Game/Serial Ports V1.41

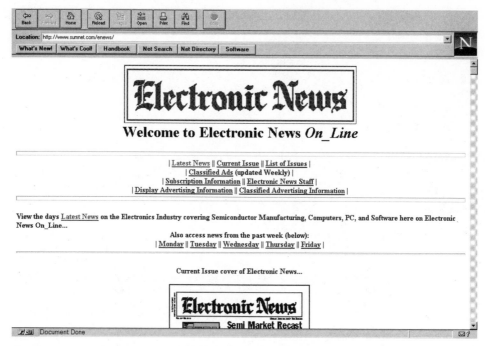

Figure 3-95   On-line news.

Parts

See Components.

PC Articles

http://performance.netlib.org/
performance/html/spec.html
SPEC Benchmark Information

http://www.mordor.com/aolsz/docs/
pcdead.html
The Dead PC

http://www.zdnet.com/home/filters/
toc.html
ZiffNet PC Magazines and Articles

PC Hardware FAQs

ftp://bode.ee.ualberta.ca/pub/
cookbook/

ftp://rtfm.mit.edu/pub/usenet/
news.answers/pc-hardware-faq

http://www.cis.ohio-state.edu/
hypertext/faq/usenet/pc-hardware-faq/
top.html
USENET FAQs: PC Hardware FAQs

http://www.paranoia.com:80/~filipg/
HTML/LINK/Fils_FAQ.html
Fil's FAQ-Link-In Corner V2.01

http://www.paranoia.com/~filipg/
HTML/LINK/F_LCD_menu.html
LCD FAQ

http://www.qucis.queensu.ca/home/
pham/homebuilt/achh.faq.html
alt.comp.hardware.homebuilt FAQ

PC Lube and Tune

http://pclt.cis.yale.edu/pclt/
default.htm

PC Magazine Search Engine

http://www.zdnet.com/~pcmag/
search.htm

PCMCIA Information

http://www.blackbox.com/bb/
refer/mobile/pcmcia/p3.html/
tig1d60
Understanding PCMCIA

PC Repair Notes

http://www.paranoia.com/~filipg/
HTML/FAQ/BODY/Repair.html
sci.electronics.repair FAQ

PC Setup Procedures

ftp://ftp.firmware.com/ftp/utils/
setup.exe

Ports

See COM Ports and IRQs.
See Parallel Port Documents.
See Serial Communication Information.

Professional Organizations

http://info.isoc.org/
The Internet Society

http://infoventures.com/
EMF Link

http://www.aeanet.org/
AEA - American Electronics Association

http://www.computer.org/
IEEE Computer Society

http://www.copper.org/
The Copper Industry Page

http://www.eff.org/
Electronic Frontier Foundation

http://www.eia.org/cema/
CEMA - Consumer Electronics Manufacturers Association

http://www.emclab.umr.edu/aces/
ACES - Applied Computational Electromagnetics Society

http://www.engineers.com/
The Engineer's Club

http://www.englib.cornell.edu/ice/
ice-index.html
ICE - Internet Connections for Engineering

http://www.epri.com/
The Electric Power Research Institute

http://www.hike.te.chiba-u.ac.jp/
ikeda/IEC/home.html
IEC - International Electrotechnical Commission

http://www.iee.org.uk/
IEE - Institution of Electrical Engineers [UK]

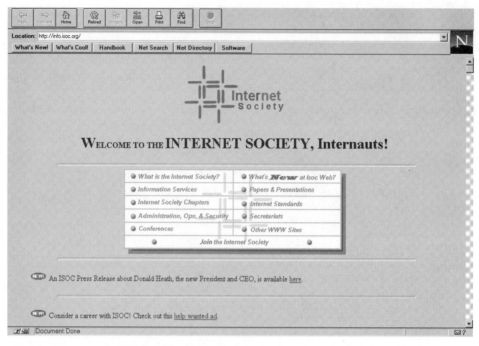

Figure 3-96   Professional organization.

ftp://ftp.ieee.org/

gopher://gopher.ieee.org/

http://www.ieee.org/
IEEE - Institute of Electrical and Electronic Engineers

http://www.iso.ch/welcome.html
ISO - International Standards Organization

http://www.itu.ch/
ITU - International Telecommunication Union [formerly CCITT]

http://www.lmsc.lockheed.com/aiaa/sf/links.html
AIAA - American Institute of Aeronautics and Astronautics

http://www.oakridge.com/aama/factsheet.html
AAMA - Asian-American Manufacturers Association

http://www.nist.gov/welcome.html
NIST - National Institute of Standards and Technology

http://www.specbench.org/
SPEC - Standard Performance Evaluation Corp.

http://www.spie.org/
SPIE - International Society of Optical Engineers

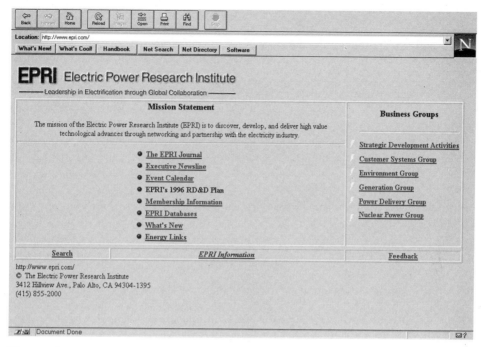

Figure 3-97 Professional organization.

Prototyping/PCB Information

http://engr-www.unl.edu/ee/eeshop/proto.html
Electronic Prototyping: Tips and Pitfalls

http://www.apcircuits.com/glosdef.htm
PCB Glossary

http://www.databahn.net/greenckt/
Green CirKit

http://www.starcon.com/bririch/html/brch1002.htm
Technology Roadmap - Designer's Hotlist

Robotics Information

ftp://rtfm.mit.edu/pub/usenet-by-hierarchy/comp/robotics/

http://piglet.cs.umass.edu:4321/robotics.html
Robotics Internet Resources Page

http://robotics.com/index.html
Arrick Robotics Page

http://www.eskimo.com/~zchris/index.html
Zorin Microcontroller Products

http://www.frc.ri.cmu.edu/robotics-faq/
Robotics Frequently Asked Questions List

S3 Video Drivers

ftp://micros.hensa.ac.uk/mirrors/cica/win3/drivers/video/s3-24b8.zip

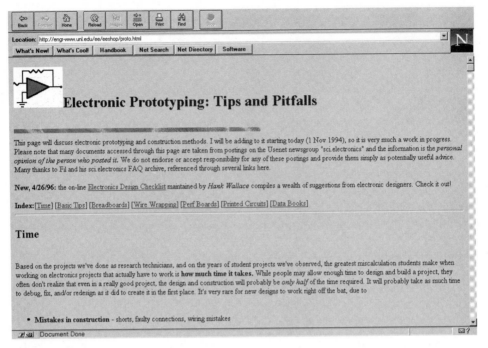

Figure 3-98 Prototyping contact.

Science Hobby Sites

ftp://rtfm.mit.edu/pub/
usenet-by-hierarchy/sci/electronics/

http://www.AutomationNET.com/
Automation NET

http://www.eskimo.com/~billb/
amasci.html
Amateur Science

http://www.paranoia.com:80/~filipg/
HTML/sci_electronics_FAQ.html
sci.electronics FAQ

http://www.wenzel.com/mystrylb.htm
Wenzel Associates Technical Library

Screen Saver Software Sites

http://www.sdust.com/
Stardust Software Page

http://www.sirius.com/~ratloaf/
Screen Savers A-Z

SCSI Information

http://www.cis.ohio-state.edu/
hypertext/faq/usenet/scsi-faq/
top.html

Search Engines and Reference Sites

http://altavista.digital.com/
Digital's Web Search

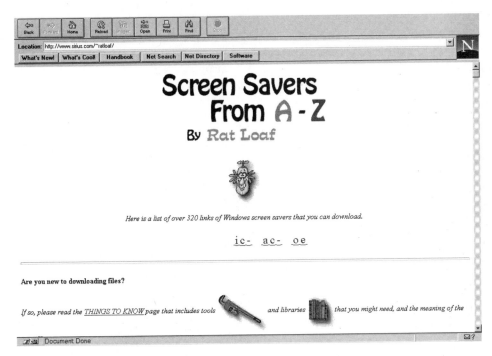

Figure 3-99   Software site.

http://galaxy.einet.net/galaxy/
Engineering-and-Technology/
Electrical-Engineering.html
Galaxy List of EE Sites

http://gnn.com/wic/wics/index.html
Whole Internet Catalog [GNN Select]

http://netcenter.com/yellows.html
Interactive URL Yellow Pages

http://src.doc.ic.ac.uk/bySubject/
Computing/Overview.html
Computing Library

http://www.bdi.ie/bdi/directs/1.htm
*Business Directory International -
Instrumentation and Controls*

http://www.cedar.buffalo.edu/
adserv.html
Cedar National Address Server

http://www.cistron.nl/~nctnico/
manulist.htm
Company list on Hardware Web

http://www.collegenet.com/
CollegeNET

http://www.ctrl-c.liu.se/other/
admittansen/netinfo.html
Electronic Hotlinks

http://www.infras.com/semico.htm
*INFRASTRUCTURE Semiconductor
Index*

http://www.micropat.com/
Patent Search site

http://www.paranoia.com/~filipg/
HTML/FAQ/BODY/F_Magazines.html
Electronics & Science Oriented Magazines V3.01

http://www.prosonline.com/proshome/
comptec.htm
Pros Online

http://www.scescape.com/worldlibrary/
business/companies/elec.html
The World Library

http://www.techexpo.com/
academia.html
TechExpo DIRECTORY

http://www.techexpo.com/directry.html
Directory of Corporations

http://www.techexpo.com/events/
evnts-p1.html
Technical Conferences

http://www.techexpo.com/
gov_data.html
Government Technical Sites

http://www.techexpo.com/
tech_mag.html
Directory of Technical Magazines

http://www.techexpo.com/
tech_soc.html
Directory of Societies & Organizations

http://www.tollfree.att.net/dir800/
AT&T "800" Directory

http://www.vts.com/
Virtual Electronics Trade Show

http://www.webscope.com/elx/
homepage.html
Electronics Manufacturers

Serial Communication Information

http://web.aimnet.com/~jnavas/
modem/faq.html
Navas 28800 Modem FAQ

Setup

See PC Setup Procedures.

Simulation

ftp://ftp.fsrv.atv.tuwien.ac.at/
Intusoft ICAP Demo Simulator

ftp://ftp.funet.fi/pub/msdos/Simtel/
electric/
Spice software

http://Dcpu1.cs.york.ac.uk:6666/
fisher/mkfilter/
Interactive Filter Design

http://engr-www.unl.edu/ee/eeshop/
cad.html
Electronic Design Software

http://picea.hut.fi/aplac/tutorial/
main.html
APLAC Tutorial

http://teal.ece.ucdavis.edu/sscrl/clcfaq/f
aq/faq-toc.html
comp.lsi.cad / comp.lsi FAQ

http://www.bahnhof.se/~visionics/
Visionics EDA

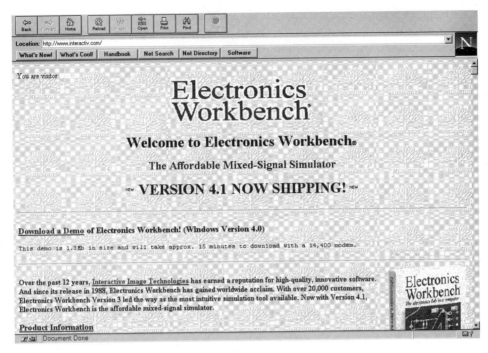

Figure 3-100 Simulation contact.

http://www.interactiv.com/
Electronics Workbench

Software Archives and Sites

ftp://ftp.syd.dit.csiro.au/pub/ken/
zcc096.zip
Small C for the Z80

ftp://oak.oakland.edu/SimTel/win3/
educate/

http://execpc.com/~wmhogg/
share.html
The ShareWare Resource Page

http://hertz.njit.edu/~rxy5310/
picb.html
PicBlaster PIC Programmer

http://home.ptd.net/~dkt/win95/
win95.htm
*Superior Software for Internet Telecom-
munications Enthusiasts*

http://udgftp.cencar.udg.mx/ingles/
tutor/Assembler.html
Assembly Language Tutorial

http://users.aol.com/ericb98398/
index.html
Personal MicroCosm

http://weber.u.washington.edu/
~pfloyd/ee/programs/index.html
Programs in Circuits Archive

http://www.acs.oakland.edu/
cgi-bin/vsl-front/
VSL front desk

http://www.acs.oakland.edu/oak/
OAK Software Repository

http://www.coast.net/SimTel/
SimTel Software Repository

http://www.csusm.edu/cwis/
winworld/winworld.html
CSU Windows Shareware Archive

http://www.jumbo.com/
Jumbo Shareware Site

http://www.shareware.com/SW/NFF/
0,1,,05040100.html
Find Software on the Internet

Software Engineering Information and Links

http://rbse.jsc.nasa.gov/virt-lib/
soft-eng.html
Software Engineering Virtual Library

Software Reviews (See Chapter 4)

http://www.ucc.uconn.edu/
~wwwpcse/wcool.html

Sybase FAQ Archives

http://www.acs.ncsu.edu/Sybase/

Technology News

See News Contacts and Sites.

Television and Antenna Information

ftp://sparky.cs.yale.edu/pub/video/

Universities (EE Departments, Organizations, and Museums)

http://snf.stanford.edu/
ComputationalPrototyping/
Stanford Computational Prototyping

http://cra.org/
Computing Research Association

http://dspserv.eng.umd.edu/
DSP Lab

http://kabuki.eecs.berkeley.edu/
Research in Analog IC Design

http://lol.cs.columbia.edu/~library/
Technical Report Archive

http://ocswebhost.colorado.edu/
Optoelectronic Computing Systems Center

http://cis.wustl.edu/
Center for Imaging Science

http://sun158.dees.unict.it/index.html
Systems and Control Group Home Page

http://www-bsac.eecs.berkeley.edu/
Berkeley Sensor & Actuator Center

http://www.ccsm.uinc.edu/ccsm/
Semiconductor Center

http://snf.stanford.edu/cis/
Stanford Center for Integrated Systems

http://www-cis.stanford.edu/NanoNet/
NanoNet Information Central

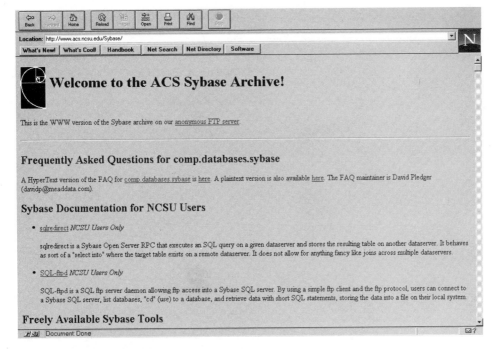

Figure 3-101 Sybase archive.

http://www.eecs.umich.edu/dp-group/
University of Michigan: Department of Electrical Engineering and Computer Science

http://www.ee.iitb.ernet.in/
EE Department: Indian Institute of Technology, Bombay

http://www-ee.stanford.edu/
ee/tcad/programs.html
Stanford TCAD Tools

http://www.eng.auburn.edu/
department/ee/amstc/amstc.html
AMSTC (Alabama Microelectronics Science and Technology Center) Home Page

http://www.engr.wisc.edu/centers/
ercpam/ercpan.html
Engineering Research Center for Plasma-Aided Manufacturing

http://www.mrc.uidaho.edu:80/
Microelectronics Research Center (MRC)

http://www-mtl.mit.edu/
MIT Microsystems Technology Laboratories

http://www-sitn.stanford.edu/sitn.html
Stanford Instructional Television Network

http://www.shef.ac.uk/~eee/esg/
Sheffield, UK: The Electronic Systems Group

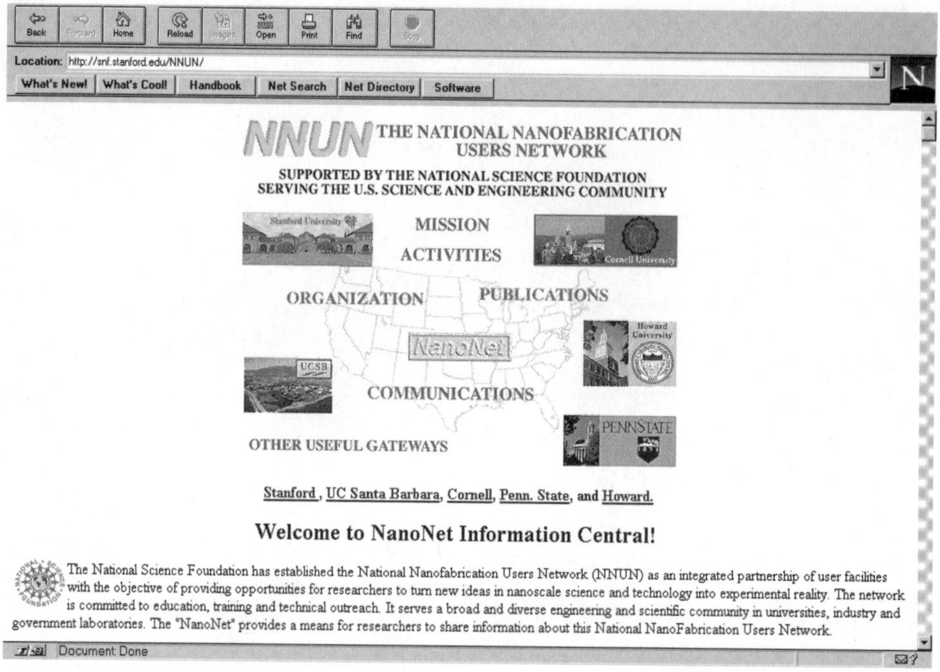

Figure 3-102   University resource.

http://www.thetech.org/
San Jose Museum of Innovation

US Postal Service (good reference for small businesses)

http://www.usps.gov/

Utility Software

http://www.bae.ncsu.edu/bae/people/faculty/walker/hotlist/graphics.html
Graphics viewers, editors, utilities and information

CR Troubleshooting and Information

http://bradley.bradley.edu/~fil/vcr_html/vcrindex.html
Index of VCR Q&A Questions

http://www.magnavox.com/electreference/videohandbook/vcrs.html
The Video Cassette Recorder Electronics Reference

http://www.paranoia.com:80/~filipg/HTML/LINK/F_MacroVision.html
MACROVISION FAQ v1.0c

VESA Information

http://www.vesa.org/

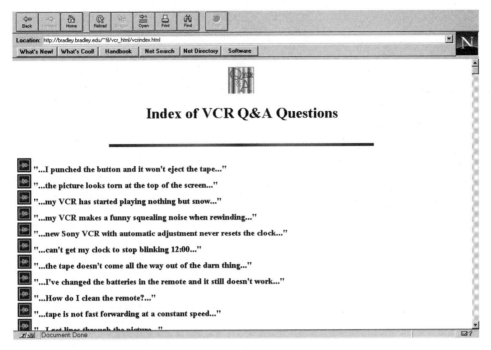

Figure 3-103   VCR contact.

Video Drivers

http://www.ege.edu.tr:80/cica/
index/drivers/video.html

Virus Scanners and Information

ftp://ftp.srv.ualberta.ca/pub/dos/
virus/

http://www.umcc.umich.edu/~doug/
virus-faq.html
Virus FAQ

VME Bus Information

http://www.ee.ualberta.ca/archive/
vmefaq.html
comp.arch.bus.vmebus FAQ

http://www.vita.com/
VMEbus International Trade Association

Web Books - On-line Publishing

http://www.opennet.com/webbooks/

Weekly Bookmark (See Chapter 4)

http://www.webcom.com/weekly/
wb/weekly.html
The Weekly Bookmark

Windows/Windows 95 Articles, Software Archives, and Related Sites

http://biology.queensu.ca/~jonesp/
OmniCon 96 - The Best Unit Conversion Utility for Windows 95 and Windows NT!

http://commline.com/falken/
tools.shtml
*Falken's Cyberspace Tools (lots of share-
ware)*

http://coyote.csusm.edu/cwis/
winworld/winworld.html
*California State University Windows
Shareware Archive*

http://www.dylan95.com
*Dylan Greene's Windows 95 Starting
Pages*

http://cws.wilmington.net/
*Stroud's Consumate Winsock Applica-
tions*

http://home.ptd.net/~arrow/
software.htm
Windows 95 Software and Applications

http://miso.wwa.com/~tyocum/
The Windows 95 Start Page

http://ourworld.compuserve.com/
homepages/G_Carter/
Get Help With Windows 95

http://sage.cc.purdue.edu/~xniu/
winsock/win95.htm
*Xiaomu's List of Windows 95 Internet
Applications*

http://techweb.cmp.com/nc/607/
607sneak1.html
CMP's Windows 95 Preview Pages

http://users.aol.com/scottb4197/
win95tip.htm
Windows 95 Tips

http://walden.mo.net/~devino/
Look in at Windows 95

http://walden.mo.net/~rymabry/
index.html
Robin's Nest

http://www.aa.net/~pcd/
Windows 95 documents

http://www.access.digex.net/~rfowler/
software.html
Ken Fowler's Home Page

http://www.avenue.com/forum/
win95.html
Microsoft Slackville

http://www.bhs.com/application.center/
The Windows NT Application Center

http://www.bluesquirrel.com/free.htm
Blue Squirrel Software

http://www.bond.edu.au/gnn/wic/
windows.01.html
Windows Shareware Archives

http://www.contrails.com/knapen/
win95.htm
Site with Windows 95 Tips

http://www.creativelement.com/
win95ann/
Windows 95 Annoyances

http://www.cris.com/~randybrg/
win95.html
Randy's Windows 95 Resource Center

http://www.generation.net/~maflalo/
big/
Internet Starting Point for Windows95

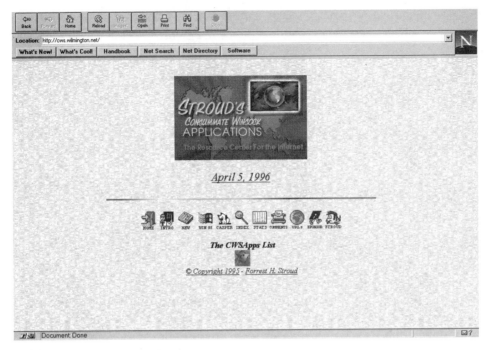

Figure 3-104   Winsock resource.

http://www.delta.edu/~rlhoward/
Bud's Home Page

http://www.eagle.ca/~jdi/johnfinal.html
*John Irvine's Windows95 Unofficial
Home Page*

http://www.euronet.nl/users/ries/
w95links.html
Windows 95 Links

http://www.execpc.com/~kjking/
index.html
KJ King's Windows 95 Web Page

http://www.executrain.com/
ExecuTrain Windows 95 Training

http://www.gov.nb.ca/hotlist/
win95.htm
Extensive Windows 95 Links

http://www.helpwin95.com/
Internet Presence Providers

http://www.cfanet.com/bmoore/
Brian's Windows 95 Web Page

http://www.jagat.com/christopher/
chris_w95.html
*The Most Needed Add-ons for Windows
95*

http://www-leland.stanford.edu/
~llurch/win95netbugs/faq.html
Windows 95 Networking FAQ

http://www.mcs.net/~jlueders/
windows95.html
John's Cool Windows 95 Page

http://www.netex.net/w95/
windows95/flute.zip
Windows 95 Live Software Archive

http://www.NetEx.net:80/w95/
Windows 95 Expert Forum

http://www.nidlink.com/~hutch/
win95.html
Hutch's Windows 95 Page

http://www.pacificrim.net/~bugnet/
BugNet Online

http://www.peinet.pe.ca:2080/
Chorus/U_report/urep.html
The Windows Utility Report '96

http://www.planet-hawaii.com/
global/win95.html
List of Windows 95 Links

http://www.pris.bc.ca/gunman/
Gunman's Win95 Headquarters

http://www.process.com/win95/
win95ftp.htm
Process Software's Windows 95 Archive

http://windows.rust.net/w95tips.html
Frank's Bag 'O Tips

http://www.scott.net:80/~gtaylor/
win95.html
House of November: Windows 95 Site

http://www.shasta-co.k12.ca.us/www/
windows95/links.html
Windows 95 Links

http://www.southwind.net/~leeb/
95slip.html
*Windows 95 SLIP Support Installation
[article]*

http://www.southwind.net/~leeb/
win95.html
*The (Unofficial) Windows 95 Home
Page*

http://www.tucows.com
Tucows Collection of Winsock Software

http://www.usa.net/~rduffy/
registry.htm
*Getting the Most from the Windows 95
Registry [article]*

http://www.w95net.com/
Windows 95 Net

http://www.webcom.com/~hywan/
bcheong/win95/win95.html
BenZ Windows 95 Page

http://www.whidbey.net/~mdixon/
qaid0001.htm
Mike Dixon's Windows 95 QAID

http://www.win95.com/
Windows 95 Reference Site

http://www.windowatch.com/
WindoWatch Home Page

http://www.wineasy.se/bjornt/
win95.html
Bjorns Windows 95 Page

http://www.winsite.com/
WinSite

http://www.winternet.com/
~rgraner/win95.html
Windows 95 Information Links

http://www2.csn.net/~medlin/
Pink Elephant Windows Site

4

Mail Lists and News Groups

NEWS GROUPS

Computers

alt.binaries.pictures.fine-art.graphics
Art created on computers (moderated list)

alt.business.import-export.computer
International trade in computers

alt.comp.acad-freedom.news
Academic freedom issues related to computers (moderated list)

alt.comp.acad-freedom.talk
Academic freedom issues related to computers

alt.comp.tandem-users
Users of Tandem computers

alt.computer.consultants
The business of consulting about computers

alt.folklore.computers
Stories and anecdotes about computers (some true!)

alt.solaris.x86
Sun's Solaris on Intel x86 compatible computers

alt.sources.mac
Source code for Apple Macintosh computers

alt.sys.icl
International Computers Limited hardware and software

aus.ads.forsale.computers
Private "for sale" ads for computers

aus.computers
Miscellaneous computer topics

aus.computers.ai
Discussion of artificial intelligence

aus.computers.amiga
Discussions about Commodore Amiga computers

aus.computers.cdrom
Discussion of CD-ROMs and computers

aus.computers.ibm-pc
Discussions about the IBM PC & clones

aus.computers.linux
Discussion of Linux and computers

aus.computers.logic-prog
Logic and programming topics about computers

aus.computers.mac
Discussion of Mac computer systems

aus.computers.os2
Discussion of OS/2 on the PC

aus.computers.parallel
Parallel processing topics

aus.computers.sun
A discussion of Sun computer hardware

av.computer
Computers in the Antelope Valley

ba.market.computers
For Sale/Wanted: Computers and software

be.comp
Computers in Belgium

bit.listserv.mbu-l
Megabyte University-Computers and Writing

biz.marketplace.computers.discussion
Discussion of computer merchandising

biz.marketplace.computers.mac
Macintosh hardware/software

biz.marketplace.computers.other
Other computer hardware/software

biz.marketplace.computers.pc-clone
PC-compatible hardware/software

biz.marketplace.computers.workstation
Computer workstation hardware/software

biz.marketplace.services.computers
Computer services offered/wanted

chile.soft-l
Software for microcomputers in Chile

clari.tw.computers.apple
Apple corporate and product news (moderated list)

clari.tw.computers.apple.releases
Releases about Apple and its products (moderated list)

clari.tw.computers.cbd
Data processing and telecommunication (moderated list)

clari.tw.computers.entertainment.releases
Releases regarding computer entertainment (moderated list)

clari.tw.computers.in_use
Using computers in industry and education (moderated list)

clari.tw.computers.industry_news
News of the computer industry (moderated list)

clari.tw.computers.misc
Miscellaneous computer news (moderated list)

clari.tw.computers.networking
Computer networking products (moderated list)

clari.tw.computers.networking.releases
Releases regarding computer networking (moderated list)

clari.tw.computers.pc.hardware
Intel-based computer hardware (moderated list)

clari.tw.computers.pc.hardware.releases
Releases regarding Intel-based hardware (moderated list)

clari.tw.computers.pc.software
A discussion of MS-DOS & Windows software (moderated list)

clari.tw.computers.pc.software.releases
Releases regarding Windows software (moderated list)

clari.tw.computers.peripherals.releases
Releases regarding computer peripherals (moderated list)

clari.tw.computers.releases
Releases regarding miscellaneous computer news (moderated list)

clari.tw.computers.retail.releases
Releases about the retail computer industry (moderated list)

clari.tw.computers.unix
News of UNIX and similar operating systems (moderated list)

clari.tw.computers.unix.releases
Press releases covering UNIX (moderated list)

clari.tw.features
Features about technical industries and computers (moderated list)

comp.ai.nat-lang
Natural language processing by computers

comp.arch.arithmetic
Implementing arithmetic on computers/digital systems

comp.binaries.psion
Binary files for the range of Psion computers (moderated list)

comp.dsp
Digital Signal Processing using computers

comp.misc
General topics about computers not covered elsewhere

comp.music.midi
Computers as components in MIDI music systems

comp.music.misc
Miscellaneous use of computers in music

comp.music.research
The use of computers in music research and composition (moderated list)

comp.os.vms
DEC's VAX line of computers & VMS

comp.protocols.tcp-ip.ibmpc
TCP/IP for IBM-compatible personal computers

comp.risks
Risks to the public from computers and users (moderated lists)

comp.security.misc
Security issues of computers and networks

comp.sys.acorn.advocacy
List advocating Acorn computers and programs

comp.sys.acorn.extra-cpu
Discussion of extra CPUs in Acorn computers

comp.sys.acorn.networking
The networking of Acorn computers

comp.sys.acorn.programmer
Programming of Acorn computers

comp.sys.alliant
Information and discussion about Alliant computers

comp.sys.atari.advocacy
Discussing the pros and cons of Atari computers

comp.sys.att
Discussions about AT&T microcomputers

comp.sys.cdc
Discussion of Control Data Corporation computers (i.e., Cybers)

comp.sys.concurrent
The Concurrent/Masscomp line of computers (moderated lists)

comp.sys.encore
Encore's MultiMax computers

comp.sys.handhelds
Handheld computers and programmable calculators

comp.sys.hp.hpux
Issues pertaining to HP-UX & 9000
series computers

comp.sys.hp.mpe
Issues pertaining to MPE & 3000
series computers

comp.sys.ibm.pc.misc
Discussion about IBM personal computers

comp.sys.laptops
Laptop (portable) computers

comp.sys.m88k
Discussion about 88k-based computers

comp.sys.misc
Discussion about computers of all
kinds

comp.sys.ncr
Discussion about NCR computers

comp.sys.next.hardware
Discussing the physical aspects of
NeXT computers

comp.sys.oric
Oric computers (Oric1, Atmos,
Telestrat, et cetera)

comp.sys.pen
Interacting with computers through
pen gestures

comp.sys.psion
Discussion about PSION Personal
Computers and Organizers

comp.sys.pyramid
Pyramid 90x computers

comp.sys.ridge
Ridge 32 computers and ROS

comp.sys.sgi.hardware
Base systems and peripherals for Iris
computers

comp.sys.sinclair
Sinclair computers (i.e., the ZX81,
Spectrum and QL)

comp.sys.super
Supercomputers

comp.sys.tahoe
CCI 6/32, Harris HCX/7, & Sperry
7000 computers

comp.sys.tandy
Discussion about Tandy computers:
new and old

comp.sys.zenith.z100
The Zenith Z-100 (Heath H-100)
family of computers

comp.unix.aux
The version of UNIX for Apple Macintosh II computers

comp.unix.cray
Cray computers and their operating
systems

dc.forsale.computers
Computer equipment and computer
peripherals

de.comp.standards
Computerstandards und ihre Auswirk-
ungen (Computer Standards: German)

de.rec.games.computer
Diskussionen rund um Computer-
spiele (Computer Games: German)

fido7.bank-technology
Computer technology in banks

fido7.bank_support
Computer support in banks

fido7.dec
A discussion of DEC computers

fido7.zx_spectrum
ZX Spectrum computers

fj.comp.misc
General topics about computers not
covered elsewhere

fj.comp.music
Topics about computers and music

fj.news.group.comp
About newsgroups for discussions of
computers

fj.sys.ibmpc
Discussion about IBM Personal Computers and clones

fj.sys.j3100
Discussion about TOSHIBA J3100-
family computers

fj.sys.misc
Discussion about computers of all
other kinds

fj.sys.newton
Discussion on computers equipped
with Newton technology

fj.sys.pc98
Discussion about NEC PC-9800 and
other computers

fj.sys.x68000
Discussion about Sharp X68000 and
other computers

fl.comp
General computers in Florida

han.comp.hangul
How Korean Hangul can be used in
computers

han.sys.hp
Hewlett-Packard computers, HP-UX.

iijnet.sys.ibm-pc
Discussions about IBM Personal computers and clones

info.solbourne
Discussions and information about Sol-
bourne computers (info zsolbourne
@acsu.buffalo.edu) (moderated)

misc.forsale.computers.discussion
Discussions only about items for sale

*misc.forsale.computers.mac-
specific.cards.misc*
A discussion of Macintosh expansion
cards

*misc.forsale.computers.mac-
specific.cards.video*
A discussion of Macintosh video cards

misc.forsale.computers.mac-specific.misc
A discussion of other Macintosh equipment

misc.forsale.computers.mac-specific. portables
Discussion list for portable Macintosh systems

misc.forsale.computers.mac-specific.software
A discussion of Macintosh software

misc.forsale.computers.mac-specific.systems
Complete Macintosh systems

misc.forsale.computers.memory
Memory chips and modules for sale and wanted

misc.forsale.computers.modems
Modems for sale and wanted

misc.forsale.computers.monitors
Monitors and displays for sale and wanted

misc.forsale.computers.net-hardware
Networking hardware for sale and wanted

misc.forsale.computers.other.misc
Miscellaneous other equipment

misc.forsale.computers.other.software
Software for other systems

misc.forsale.computers.other.systems
Other complete types of systems for sale

misc.forsale.computers.pc-specific.audio
PC audio equipment

misc.forsale.computers.pc-specific.cards.misc
PC expansion cards

misc.forsale.computers.pc-specific.cards.video
PC video cards

misc.forsale.computers.pc-specific.misc
Other PC-specific equipment

misc.forsale.computers.pc-specific.motherboards
PC motherboards

misc.forsale.computers.pc-specific.portables
Portable PC systems

misc.forsale.computers.pc-specific.software
PC software

misc.forsale.computers.pc-specific.systems
Complete PC systems

misc.forsale.computers.printers
Printers and plotters for sale and wanted

misc.forsale.computers.storage
Disk, CDROM and tape drives for sale and wanted

misc.forsale.computers.workstation
Workstation-related computer items for sale or wanted

misc.kids.computer
Discussion of the use of computers by children

nersc.sas
Issues pertaining to the Supercomputing Auxiliary Service computers at NERSC

nj.market.computers
For Sale/Wanted: Computers and software

nl.comp
Discussies over computers etc.

Engineering

alt.engineering.electrical
Discussing electrical engineering

aus.aswec
Australian Software Engineering Conference

can.schoolnet.eng.jr
SchoolNet Engineering for elementary students

can.schoolnet.eng.sr
SchoolNet Engineering for high school students

comp.software-eng
Software engineering and related topics

cu.courses.ecen5593
CU Electrical Engineering Course 5593

cu.courses.geen1400-010
CU General Engineering Course 1400 Section 010

cu.courses.geen1400-030
CU General Engineering Course 1400 Section 030

fj.engr.control
Topics about control system engineering

fj.engr.elec
Electrical and electronic engineering

fj.engr.misc
General topics about engineering

gwu.seas.cs219
GWU School of Engineering-Computer Science class

info.ietf
Internet Engineering Task Force (IETF) discussions (ietf@venera.isi.edu) (moderated)

info.wisenet
Women in Science and Engineering NETwork (wisenet@uicvm.uic.edu) (moderated)

mit.eecs.discuss
Electrical Engineering at Computer Science at MIT

muc.lists.ietf
Internet Engineering Task Force

nctu.cm.general
NCTU Dept of Communication Engineering

nctu.ee.general
NCTU Dept of Electronics Engineering

oh.acad-sci
OAS: science, engineering, technology and education

sci.engr
Technical discussions about engineering tasks

sci.engr.control
The engineering of control systems

su.class.e235
Space Systems Engineering

su.class.ees231b
EES 231B Decision Engineering

ucb.class.engin260
Engineering 260

ucb.class.engr190
Engineering 190

ucb.ee.grads
Discussion group for Electrical Engineering graduates

ucb.org.ses
Group for the Society of Engineering Science

uiuc.org.cse
Seminar series in Computational Science and Engineering

uiuc.org.engcouncil
UIUC's Engineering Council

uiuc.org.eoh
Engineering Open House discussions

uiuc.org.eoh.announce
Announcements for Engineering Open House

uiuc.org.eoh.design
Discussion of Engineering Open House design contest

uiuc.org.eoh.general
General discussion concerning Engineering Open House

ukr.comp.hardware
Hardware problems of computer engineering

umich.eecs.announce
UofM Electrical Engineering and Computer Science Announcements

umich.eecs.ce.students
UofM EECS Computer Engineering Discussions

umich.org.umec
Newsgroup for the University of Michigan Engineering Council

umn.cs.class.5180
Software engineering

umn.cs.class.5181
"Software Engineering II" at the University of Minnesota

ut.ecf.comp9t5
Playpen for Computer Engineering graduating class

ut.ecf.student
Discussion group for University of Tennessee engineering students

ut.engineering.general
Faculty of Applied Science and Engineering news and discussion

ut.engineering.seminars
Faculty of Applied Science and Engineering seminar announcements

utexas.class.cs373
CS373 Software Engineering

uw.cs.eee
Computer Science / Electrical Engineering

uw.ee.grad
Electrical Engineering

uwo.csd.cs201
Software Engineering, Algorithms and Data Structures Course

wu.cec.general
Informational postings by the Center for Engineering Computing (moderated list)

Computer and Electronics Science

alt.comp.hardware.homebuilt
Discussion for home-brew PC hardware builders

alt.hackers
A general "computer hackers" discussion group

comp.dcom.telecom
Telecommunications discussion group

comp.dsp
Discussion of digital signal processing

comp.home.automation
Discussion of home automation equipment

comp.realtime
Discussion of real-time computing issues

comp.robotics
Hobbyist robotics equipment and techniques

comp.sys.intel
Discussion of Intel processors and Intel ICs

sci.electronics
General electronic discussions and information

sci.electronics.basic
Introductory discussions about electronics

sci.electronics.cad
PCB design and capture, and other CAD topics

sci.electronics.components
Discussion group covering electronics parts

sci.electronics.design
A group discussing essential electronic design

sci.electronics.equipment
A discussion group for test equipment

sci.electronics.misc
A group discussing various electronics issues

sci.electronics.repair
Discussing electronics troubleshooting/repair

sci.engr.advanced-tv
High-definition television (HDTV) discussion

sci.virtual-worlds
A general VR (virtual reality) discussion group

sci.virtual-worlds.apps
Discussion of the applications of virtual reality

Operating Systems and Programming

alt.lang.basic
BASIC language programming topics

alt.msdos.programmer
Discussion group for MS-DOS programmers

aus.computers.os2
Australian OS/2 discussion group

comp.binaries.os2
General OS/2 discussion group

comp.lang.ada
Discussion group for Ada programming

comp.lang.basic.misc
Discussion of general BASIC topics

comp.lang.basic.visual
Discussion of Visual BASIC

comp.lang.c
Discussion group for C programming

comp.lang.c++
Discussion group for C++ programming

comp.lang.forth
Discussion group for Forth programming

comp.lang.lisp
Discussion group for Lisp programming

comp.lang.modula2
Discussion group for Modula2 programming

comp.lang.pascal
Discussion group for Pascal programming

comp.lang.postscript
Discussion group for Postscript programming

comp.lang.prolog
Discussion group for Prolog programming

comp.os.ms-windows.apps.comm
Discussion of Windows communication applications

comp.os.ms-windows.apps. compatibility.win95
Discussion of Windows 95 compatibility

comp.os.ms-windows.apps.misc
Discussion of various Windows applications

comp.os.ms-windows.apps.utilities
Discussion of various Windows utilities

comp.os.ms-windows.apps.utilities.win95
Discussion of various Windows 95 utilities

comp.os.ms-windows.apps.winsock.misc
Discussion of various winsock topics

comp.os.ms-windows.apps.winsock.news
Discussion of Windows winsock news

comp.os.ms-windows.apps.winsock.mail
Discussion of Windows winsock mail

comp.os.ms-windows.misc
Various Windows discussion topics

comp.os.ms-windows.networking.tcp-ip
Networking with Windows using
TCP-IP

comp.os.ms-windows.networking.win95
General networking topics under
Windows 95

comp.os.ms-windows.networking.windows
General networking topics under
Windows

comp.os.ms-windows.programmer.controls
Discussion group for Windows controls

comp.os.ms-windows.programmer.drivers
Discussion group for creating Windows drivers

comp.os.ms-windows.programmer.graphics
Discussion group for Windows
graphics

comp.os.ms-windows.programmer.memory
Discussion group for Windows memory issues

comp.os.ms-windows.programmer.misc
Miscellaneous programming issues
under Windows

*comp.os.ms-windows.programmer.
multimedia*
Multimedia programming issues discussion

comp.os.ms-windows.programmer.networks
Network programming issues under
Windows

comp.os.ms-windows.programmer.ole
Discussion of Windows OLE programming

comp.os.ms-windows.programmer.tools
Programming development discussion

comp.os.ms-windows.programmer.winhelp
Group for the discussion of Windows
help files

comp.os.ms-windows.win95.misc
Discussion of miscellaneous Windows
95 topics

comp.os.ms-windows.win95.setup
Discussion of Windows 95 setup topics

comp.os.msdos.programmer
A discussion group for MS-DOS programmers

comp.os.os2.advocacy
Discussion advocating the use of
OS/2

comp.os.os2.announce
Forum for new OS/2 announcements

comp.os.os2.apps
Discussion of OS/2 applications

comp.os.os2.beta
Forum to discuss OS/2 beta software

comp.os.os2.bugs
Forum to discuss OS/2 bugs

comp.os.os2.games
Forum to discuss OS/2 games

comp.os.os2.misc
Discussion of various OS/2 topics

comp.os.os2.multimedia
Discussion of OS/2 multimedia topics

comp.os.os2.networking.misc
An OS/2 networking group

comp.os.os2.networking.tcp-ip
Discussion of OS/2 networks with
TCP-IP

comp.os.os2.programmer.misc
Discussion of programming general
topics

comp.os.os2.programmer.oop
Discussion of object-oriented topics

comp.os.os2.programmer.porting
Discussion of porting the OS/2 operating systems

comp.os.os2.programmer.tools
Discussion of OS/2 development tools

comp.os.os2.setup
Discussion of OS/2 setup

comp.sys.ibm.pc.programmer
General PC programming topics discussed

comp.sys.ibm.pc.soundcard.tech
Technical sound card programming
discussion

rec.games.programmer
Video game programmer's discussion
group

**Computer Business, Repair, and
Instruction**

alt.books.technical
Discussion group for technical books

alt.computer.consultants
Computer consultants discussion
group

biz.books.technical
Commercial postings of technical
books

biz.comp.hardware
Posts of computer hardware for sale

biz.comp.services
Posts of computer-related services

biz.comp.software
Posts of computer software for sale

biz.jobs.offered
Posts of available job opportunities

comp.dcom.modems
Modem discussion group

comp.os.msdos.deskview
Quarterdeck products newsgroup

comp.sys.ibm.pc.hardware.misc
PC Information

comp.sys.ibm.pc.hardware
PC FAQ

misc.books.technical
Technical book discussion group

misc.jobs.contract
Posts of available contract jobs

misc.jobs.offered
Posts of available jobs (software)

misc.jobs.resume
Posts of resumes

Radio (CB/Ham)

alt.radio.pirate
Low power AM/FM/TV pirate broadcasting

alt.radio.scanner.uk
Scanning discussions for the United Kingdom

phl.scanner
Scanner topics in the Philadelphia, PA area

rec.antiques.radio+phono
Antique radio and phono topics

rec.radio.amateur.antenna
Antennas: theory, techniques, construction

rec.radio.amateur.digital.misc
Packet and other digital radio modes

rec.radio.amateur.equipment
All about HAM radio hardware

rec.radio.amateur.homebrew
HAM construction and experimentation

rec.radio.amateur.misc
HAM radio practices, contests, events, rules, etc.

rec.radio.amateur.policy
Radio use and regulation policy

rec.radio.amateur.space
Amateur radio transmissions through space

rec.radio.broadcasting
Local area broadcast radio (moderated)

rec.radio.cb
Citizen Band discussion area

rec.radio.info
Informational postings related to radio (moderated)

rec.radio.noncomm
Topics relating to noncommercial radio

rec.radio.scanner
"Utility" broadcasting traffic above 30 MHz

rec.radio.shortwave
Shortwave radio enthusiasts

rec.radio.swap
Offers to trade and swap radio equipment

MAILING LISTS

AAASCS   (*AAAS Computer Systems*)
Mail the command info AAASCS to
LISTSERV@GWUVM.GWU.EDU

acadia-1   (*Association for Computer-Aided Design in Architecture*)

Mail the command information acadia-l to listserv@alcor.unm.edu

ACORN-L (*ACORN computers Discussion List*)
Mail the command info ACORN-L to LISTSERV@VM.EGE.EDU.TR

acscs-l (*UH Community College Computer Specialist Discussion List*)
Mail the command information acscs-l to listproc@hawaii.edu

ACSU-L (*Association of Computer Science Undergraduates*)
Mail the command information ACSU-L to listproc@cornell.edu

act-l (*Adaptive Computer Technology forum*)
Mail the command information act-l to listproc@lists.missouri.edu

agocg-info (*Advisory Group on Computer Graphics*)
Mail the command info agocg-info to majordomo@mcc.ac.uk

ami-sci (*Discussion of scientific uses of the Amiga computer*)
Mail the command info ami-sci to majordomo@phy.ucsf.edu

ami-sci-digest (*Discussion of scientific uses of the Amiga computer*)
Mail the command info ami-sci-digest to majordomo@phy.ucsf.edu

apcs (*Advanced Placement Computer Science*)
Mail the command info apcs to majordomo@acpub.duke.edu

APPL-L (*Computer applications in science and education*)
Mail the command info APPL-L to LISTSERV@VM.CC.UNI.TORUN.PL

APPLE-L (*Cornell s Apple Computer information list*)
Mail the command information APPLE-L to listproc@cornell.edu

ARJ-INFO (*General information on ARJ compression*)
Mail the command info ARJ-INFO to MAJORDOMO@DUNKEL.DE

ARRL (*Amateur Radio Relay League information*)
Mail the command help to info@arrl.org

as-computer-policy (*Communication among members of the UCI Academic Senate Computer Policy Committee.*)
Mail the command information as-computer-policy to listserv@uci.edu

AUCAP-L (*Computer Alert Project*)
Mail the command info AUCAP-L to LISTSERV@AMERICAN.EDU

auraria-campus-computers (*Auraria Campus Computers*)
Mail the command information auraria-campus-computers to listproc@carbon.cudenver.edu

AUTOCAD-L (*Using AutoCAD computer automated design CAD tools at Cornell*)
Mail the command information AUTOCAD-L to listproc@cornell.edu

AVTECH-L AV (*Repair and Service Technicians*)

Mail the command info AVTECH-L to
LISTSERV@PSUVM.PSU.EDU

AXIOM   (*AXIOM Computer Algebra
System*)
Mail the command info AXIOM to
LISTSERV@VM1.NODAK.EDU

AZCOMP   (*Arizona Computer Access
Group*)
Mail the command info AZCOMP to
LISTSERV@ASUVM.INRE.ASU.EDU

bcab   (*British Computer Association of
the Blind*)
Mail the command info bcab to
majordomo@cs.man.ac.uk

bcab-digest   (*British Computer Association of the Blind* (digest form))
Mail the command info bcab-digest to
majordomo@cs.man.ac.uk

bcs-activists   (*Boston Computer Society
activists* (volunteers))
Mail the command info bcs-activists to
majordomo@world.std.com

BESTCOM   (*Best Computer Announce-
ment List*)
Mail the command info BESTCOM to
LISTSERV@VM.EGE.EDU.TR

BICOMPAL   (*The Big Computer Pals
Discussion List*)
Mail the command info BICOMPAL to
LISTSERV@SJUVM.stjohns.edu

BLIND-L   (*Computer Use by and for the
Blind*)
Mail the command info BLIND-L to
LISTSERV@UAFSYSB.UARK.EDU

BUSETH-L   (*Business Ethics Computer
Network*)
Mail the command info BUSETH-L to
LISTSERV@UBVM.cc.buffalo.edu

C+HEALTH   (*The Health Effects of
Computer Use*)
Mail the command info C+HEALTH to
LISTSERV@IUBVM.UCS.INDIANA.EDU

CADAM-L   (*Computer Aided Design
and Manufacturing (CADAM) Interest
Group*)
Mail the command info CADAM-L to
LISTSERV@LISTSERV.SYR.EDU

CAEDS-L   (*Computer Aided Engineer-
ing Design (CAEDS) Interest Group*)
Mail the command info CAEDS-L to
LISTSERV@LISTSERV.SYR.EDU

CAEE-L   (*SEFI Working Group on Com-
puters And Engineering Education*)
Mail the command info CAEE-L to
LISTSERV@SEARN.SUNET.SE

CAI   (*Computer Assisted Instruction
Participants*)
Mail the command info CAI to
LISTSERV@vmen3.CC.EMORY.EDU

cai-l   (*For faculty using the computer
assisted instruction model*)
Mail the command information cai-l to
listproc2@bgu.edu

CAIS-L   (*Computer Algebra Informa-
tion System*)
Mail the command info CAIS-L to
LISTSERV@RZ.UNI-KARLSRUHE.DE

cal-l   (*Computer Assisted Learning List
for the College of Arts and Sciences*)

Mail the command information cal-l to
listserv@nosferatu.cas.usf.edu

CALS-CDN (*A forum for the discussion of CALS* (Computer-Aided Aquisition))
Mail the command info CALS-CDN to
LISTSERV@VM1.MCGILL.CA

CAPE-NL (*Computer Aided Process Engineering* (in Dutch; CAPE Platform))
Mail the command info CAPE-NL to
LISTSERV@HEARN.NIC.SURFNET.NL

CAPS (*Computer-Aided Prototyping System*)
Mail the command information CAPS to
listproc@wugate.wustl.edu

CARR-L (*Computer-assisted Reporting & Research*)
Mail the command info CARR-L to
LISTSERV@ULKYVM.LOUISVILLE.
EDU

CASE-L (*Computer Aided Software Engineering*)
Mail the command info CASE-L to
LISTSERV@UCCVMA.UCOP.EDU

CATIA-L (*Computer Aided Three Dimensional Interactive Applications* (CATIA))
Mail the command info CATIA-L to
LISTSERV@LISTSERV.SYR.EDU

CBH-L (*Computer Based Honors Program Discussion List*)
Mail the command info CBH-L to
LISTSERV@UA1VM.UA.EDU

CBTDG-L (*Computer Based Training Developers Group*)
Mail the command info CBTDG-L to
LISTSERV@IRLEARN.UCD.IE

cbtmm (*Discussion of computer based training, multimedia issues*)
Mail the command information cbtmm to
listproc@bgu.edu

cca (*Computer Consultants Association*)
Mail the command information cca to
listproc@ucdavis.edu

ccadvise (*SACS Computer Center Advisory Committee*)
Mail the command information ccadvise
to listserver@westga.edu

cccc (*Computer Center mailing list*)
Mail the command info cccc to
majordomo@udomo.calvin.edu

ccccstu (*Computer Center Student mailing list*)
Mail the command info ccccstu to
majordomo@udomo.calvin.edu

CCD-L (*Computer Centre Directors' List*)
Mail the command info CCD-L to
LISTSERV@ADMIN.HUMBERC.ON.CA

ccpc (*Chinese Computer Professional Club—Main list*)
Mail the command info ccpc to
majordomo@grinch.cs.buffalo.edu

ccs-news (*Campus Computer Store announcement list*)
Mail the command information ccs-news
to listproc@wugate.wustl.edu

ccs-staff (*Computer Centre Staff list*)
Mail the command information ccs-staff
to listserver@ic.ac.uk

CDROM-L (*General CD-ROM data and buying information*)
Mail the command info CDROM-L to LISTSERV@UCCVMA.UCOP.EDU

CDR-L (*Issues pertaining to the mastering of CDs*)
Mail the command index CDR-L to LISTSERV@TULSAJC.TULSA.CC.OK.US

CDROMLAN (*Using CD-ROMs on a network*)
Mail the command info CDROMLAN to LISTSERV@IDBSU.IDBSU.EDU

CDWORKS (*Discussion list for Logicraft Cdworks 2000 server*)
Mail the command info CDWORKS to CDWORKS-REQUEST@MAIL-LIST.COM

ce-train (*List for daily computer training sessions by Coop. Ext.*)
Mail the command information ce-train to listproc@ucdavis.edu

ce-trainx2 (*List for twice weekly computer training sessions by Coop. Ext.*)
Mail the command information ce-trainx2 to listproc@ucdavis.edu

cecomputer (*Cooperative Extension computer contacts discussion group*)
Mail the command information cecomputer to listproc@ucdavis.edu

CECS-L (*MU Computer Engineering and Computer Science*)
Mail the command info CECS-L to LISTSERV@MIZZOU1.MISSOURI.EDU

CEI-L (*Computer Ethics Institute List*)
Mail the command info CEI-L to LISTSERV@AMERICAN.EDU

cfs-l (*Computers For Schools*)
Mail the command information cfs-l to listproc@schoolnet.carleton.ca

CG-CHAR (*Computer Graphics Character Animation*)
Mail the command info CG-CHAR to LISTSERV@morgan.ucs.mun.ca

CGE (*Computer Graphics Education Newsletter*)
Mail the command info CGE to LISTSERV@VM.MARIST.EDU

cgu-students (*Computer Graphics Vacation Students*)
Mail the command info cgu-students to majordomo@mcc.ac.uk

CHANGE-L (*UCOP Computer Center Changes*)
Mail the command info CHANGE-L to LISTSERV@UCCVMA.UCOP.EDU

citynet-l (*CityNet—Nashville's high-speed computer network*)
Mail the command info citynet-l to majordomo@listserv.telalink.net

clics (*EC/BRA/6811 Categorical Logic in Computer Science II*)
Mail the command information clics to listproc@doc.ic.ac.uk

cmc (*CMC: Computer Mediated Communication class list*)
Mail the command information cmc to listproc@piranha.acns.nwu.edu

CMPLAW-L (*Internet and Computer Law Association*)
Mail the command info CMPLAW-L to LISTSERV@nervm.nerdc.ufl.edu

cmps4sale (*BGSU Computer Sales and Rentals*)
Mail the command information cmps4sale to listproc@listproc.bgsu.edu

CNEDUC-L (*Computer Networking Education Discussion List*)
Mail the command info CNEDUC-L to LISTSERV@TAMVM1.TAMU.EDU

CNSF-L (*Cornell National Supercomputer Facility Announcements List*)
Mail the command info CNSF-L to LISTSERV@UBVM.cc.buffalo.edu

cnsf-l (*Cornell National Supercomputer Facility Interest List*)
Mail the command information cnsf-l to listproc@listproc.wsu.edu

COCO (*COCO—Tandy Color Computer List*)
Mail the command info COCO to LISTSERV@PUCC.PRINCETON.EDU

COMCRI-L (*Computer Related Crime*)
Mail the command info COMCRI-L to LISTSERV@VM.CC.UNI.TORUN.PL

COMMODOR (*Commodore Computers Discussion*)
Mail the command info COMMODOR to LISTSERV@UBVM.cc.buffalo.edu

COMP-CEN (*Computer Center Managers' Issues*)
Mail the command info COMP-CEN to LISTSERV@UCCVMA.UCOP.EDU

COMP-TRAIN-L (*Computer Trainers List*)
Mail the command information COMP-TRAIN-L to listproc@cornell.edu

compaccess-l (*Computer Access for Individuals with Disabilities*)
Mail the command information comp-access-l to listserv@alcor.unm.edu

compharv-announce (*Computers At Harvard Announcement List*)
Mail the command info compharv-announce to majordomo@hcs.harvard.edu

complcc-l (*Computers in education at LCC*)
Mail the command information complcc-l to listproc@hawaii.edu

COMPLAW (*Legal issues related to HW, SW, comm., etc.*)
Mail a written request (with name and e-mail address) to galkin@aol.com

Comptalk (*General Computer Discussion List*)
Mail the command info Comptalk to Maiser@ccs.edge-hill-college.ac.uk

comptrain (*COMPTRAIN: Computer Training Task Force of NUDPIG*)
Mail the command information comptrain to listproc@piranha.acns.nwu.edu

COMPUNOTES (*IBM/compatibles news, reviews, web sites, etc.*)
Mail the command info COMPUNOTES to NOTES@BASIC.NET

computer-go (*Programming Computers to play Go*)

Mail the command information computer-go to listproc@anu.edu.au

computeracy   (*Joe Moxley's Computeracy mail list*)
Mail the command information computeracy to listserv@nosferatu.cas.usf.edu

COMPUTERG   (*WAMU-FM Computer Guys Announcements and Discussion List*)
Mail the command info COMPUTERG to LISTSERV@LISTSERV.AOL.COM

Computers   (*Computers for Social Change: in NYC metro area*)
Mail the command info Computers to majordomo@igc.apc.org

computersales   (*UCD BOOKSTORE Computer Shop*)
Mail the command information computersales to listproc@ucdavis.edu

COMSOC-L   (*Computers and Society ARPA Digest*)
Mail the command info COMSOC-L to LISTSERV@AMERICAN.EDU

cons   (*Student computer consultants discussion forum*)
Mail the command information cons to listproc@piranha.acns.nwu.edu

CONSULT   (*UM-St. Louis Computer Lab Consultants*)
Mail the command info CONSULT to LISTSERV@UMSLVMA.UMSL.EDU

COORDS   (*York University Computer Coordinators list*)
Mail the command info COORDS to LISTSERV@YORKU.CA

cops-l   (*Computer Operations Staff and Interested Others*)
Mail the command information cops-l to listproc@u.washington.edu

CPSR   (*Computer Professionals for Social Responsibility*)
Mail the command info CPSR to LISTSERV@GWUVM.GWU.EDU

cs-alums   (*Alumni of Dartmouth's Computer Science Department*)
Mail the command info cs-alums to majordomo@Dartmouth.EDU

cs-club   (*The Computer Science Club—A local chapter of DPMA*)
Mail the command information cs-club to listproc2@bgu.edu

cs-lab   (*Computer Systems Lab Announcements*)
Mail the command information cs-lab to listproc@cs.wisc.edu

cs-staff   (*Computer Science Staff List*)
Mail the command info cs-staff to majordomo@hub.ucsb.edu

cs-tr   (*To hear about new Computer Science Tech Reports*)
Mail the command info cs-tr to majordomo@Dartmouth.EDU

cs100   (*Computer Science 100*)
Mail the command information cs100 to listproc@callutheran.edu

CS100-DW   (*Computer Basics*)
Mail the command information CS100-DW to listproc@listproc.bgsu.edu

CS100-RC (*Computer Basics*)
Mail the command information CS100-RC to listproc@listproc.bgsu.edu

cs200 (*Computer Science 200*)
Mail the command information cs200 to listproc@callutheran.edu

cs2mf3-1 (*Discussion of Computer Science 2MF3*)
Mail the command information cs2mf3-l to listproc@informer1.CIS.McMaster.CA

cs311main (*Computer Science 311*)
Mail the command information cs311 main to listproc@callutheran.edu

cs3ca3-1 (*Discussion of Computer Science 3CA3*)
Mail the command information cs3ca3-l to listproc@informer1.CIS.McMaster.CA

csar (*Computer Sales and Rental*)
Mail the command information csar to listproc@listproc.bgsu.edu

csclub (*Computer Science Club*)
Mail the command information csclub to listproc@ucdavis.edu

cscurric (*University College Curriculum Planning for Computer Studies*)
Mail the command information cscurric to listproc@piranha.acns.nwu.edu

CSG (*Computer Study Group*)
Mail the command info CSG to LISTSERV@GWUVM.GWU.EDU

csg-list (*Computer Centre Cluster Support Group*)
Mail the command information csg-list to listserver@ic.ac.uk

CSISNEWS (*Bulletin for members of Czechoslovac Society of Computer Science*)
Mail the command info CSISNEWS to LISTSERV@EARN.CVUT.CZ

CSMS96-L (*Computer Services Management Symposium-XXIII (1996)*)
Mail the command info CSMS96-L to LISTSERV@UMSLVMA.UMSL.EDU

CSTORE (*MSU Computer Store*)
Mail the command info CSTORE to LISTSERV@MSU.EDU

cuc (*UNM Computer Use Committee*)
Mail the command information cuc to listserv@alcor.unm.edu

CUG-HELP (*Computer Users Group (CUG) Help Account*)
Mail the command info CUG-HELP to LISTSERV@UCF1VM.CC.UCF.EDU

CUG-L (*Computer Users Group (CUG)*)
Mail the command info CUG-L to LISTSERV@UCF1VM.CC.UCF.EDU

CUMREC-L (*CUMREC-L Administrative computer use*)
Mail the command info CUMREC-L to LISTSERV@VM1.NODAK.EDU

CUSSNET (*The Computer Use in Social Services Network List*)
Mail the command info CUSSNET to LISTSERV@UTARLVM1.UTA.EDU

CYBER-L (*CDC Computer Discussion*)
Mail the command info CYBER-L to
LISTSERV@uga.cc.uga.edu

CYBERSPACE-LAW (*Internet principles of law for laypeople*)
Mail the command info CYBERSPACE-LAW to LISTPROC-REQUEST@COUNSEL.COM

DE-CONF (*Committee on Computer Conferencing in Distance Education*)
Mail the command info DE-CONF to
LISTSERV@morgan.ucs.mun.ca

DELPHT-L (*Devoted to information on Borland's Delphi*)
Mail a message to hansdej@cistron.nl
with SEND help in the message subject

ECCC-L (*East Carolina University Computer Club*)
Mail the command info ECCC-L to
LISTSERV@ECUVM.CIS.ECU.EDU

ee (*Electrical and Electronic Engineering titles by C&H*)
Mail the command info ee to
majordomo@zelda.thomson.com

ED247-L (*ED247 Instructional Applications of Microcomputers*)
Mail the command information ED247-L
to listproc@cornell.edu

ed_w310 (*W310-Computer Based Education Class Discussion*)
Mail the command info ed_w310 to
majordomo@indiana.edu

elec-staff (*ANU Electronics Staff mailing list*)

Mail the command information elec-staff
to listproc@anu.edu.au

electronics (*NESDA Electronic servicers list*)
Mail the command info electronics to
majordomo@aros.net

elt150 (*Electronics technology class*)
Mail the command information elt150 to
listproc@prairienet.org

EMCDEPT (*Educational Media and Computers Department*)
Mail the command info EMCDEPT to
LISTSERV@ASUVM.INRE.ASU.EDU

ens206 (*Computer circuits class*)
Mail the command information ens206
to listproc@prairienet.org

etech (*Current Trends in Electronics Design Technology*)
Mail inquiries about etech to
anaik@cat.ernet.in

ET-W1 (*UNU/ZERI-BF Computer Conference*)
Mail the command info ET-W1 to
LISTSERV@SEARN.SUNET.SE

FACMGT (*Computer Center Facilities Management*)
Mail the command info FACMGT to
LISTSERV@ASUVM.INRE.ASU.EDU

FAIRNET (*UF Computer Fair network implementers*)
Mail the command info FAIRNET to
LISTSERV@nervm.nerdc.ufl.edu

fcr-compute (*Faculty Council on Research Computer Hardware/Software*)

Mail the command information fcr-compute to listproc@u.washington.edu

GUS-GENERAL   (*Gravis sound card information list*)
Mail the command info GUS-GENERAL to gus-general-request@apollo.COSC.GOV

hcs-compharv   (*HCS Computers @ Harvard Mailing List*)
Mail the command info hcs-compharv to majordomo@hcs.harvard.edu

HDTV   (*Issues related to developing HDTV technology*)
Mail the command info HDTV to HDTV@teletron.com

homebuilt-l   (*Discussion about Home-built Computers (NOT PC's!)*)
Mail the command info homebuilt-l to majordomo@teleport.com

HOTWIRED   (*Discussion of home-built processors*)
Mail the command info hotwired to LISTSERV@gu.uwa.edu.au

IEEE-L   (*Institute of Electrical and Electronics Engineers*)
Mail the command information IEEE-L to listproc@cornell.edu

IN-TOUCH   (*Weekly reviews of all new Win95 software*)
Mail the command subscribe IN-TOUCH to LISTSERV@PEACH.EASE.LSOFT.COM

JAVA-WIN   (*Development and use of Java and MS Windows*)

Mail the command info JAVA-WIN to majordomo@natural.com

Job-Link-Admin   (*List of new job openings or opportunities*)
Mail a message to Job-Link-Admin@listserv.job-link.com with subscribe in the subject

ksr-list   (*Discussion about the KSR family of supercomputers*)
Mail the command info ksr-list to Majordomo@hpc.uh.edu

motd   (*Message Of The Day list for fsu computers and networks*)
Mail the command information motd to listproc@mailer.fsu.edu

next-info   (*NeXT Computers at Princeton*)
Mail the command information next-info to listproc@lists.Princeton.EDU

omf-computers   (*Open discussion for OMF computer users*)
Mail the command info omf-computers to hub@XC.Org

PCTECH-L   (*Technical issues relating to the PC*)
Mail the command info PCTECH-L to LISTSERV@VM.EGE.EDU.TR

PERSON-L   (*Personal computing and maintenance issues*)
Mail the command information PERSON-L to LISTSERV@IRLEARN.UCD.IE

powelec   (*Power Electronics Seminar Mailing List*)

Mail the command info powelec to majordom@www.ee.washington.edu

SAMSUNG (*Issues pertaining to Samsung laptops*)
Mail the command information samsung to listproc@hallux.medschool.hscbklyn.edu

techscnf (*Discussion of CLSNet and computers generally*)
Mail the command info techscnf to majordomo@iclnet93.iclnet.org

TRDEV-L (*Computer training and development issues*)
Mail the command info TRDEV-L to LISTSERV@PSUVM.PSU.EDU

uci-mac (*Information and pricing of Mac Computers at UCI*)
Mail the command information uci-mac to listserv@uci.edu

WIN95-L (*Information and guidance for Windows 95*)
Mail the command sub WIN95-L to LISTSERV@PEACH.EASE.LSOFT.COM

WIN95 (*Tips, tricks, and questions regarding Win95*)
Mail the command info WIN95 to listserv@ruger.ucdavis.edu

CHAPTER

5

Resource Cross Reference

AUTOMATION

Equipment
B&B Electronics Manufacturing Co.
Robotics
Arrick Robotics

BIOS

See ICs/BIOS.

COMMUNICATIONS

Hardware
Adtran
Alpha Telecom, Inc.
AT&T
Northern Telecom
Nippon Telegraph & Telephone
Pacific Bell
Sattel
ISDN
Adtran
Alpha Telecom, Inc.
Ascend Communications

Microtronix Datacomm, Ltd.
SCii Telecom

COMPUTERS

Batteries
  Exide
  Power Express
  Sanyo
  Tauber Electronics
Connectors
  AMP, Inc.
  Aries Electronics
  Berg Electronics
  EDAC
  Major League Electronics
  Molex, Inc.
  Samtec, Inc.
Dedicated (i.e., game systems)
  3DO (The 3DO Company)
  Advanced Systems
  RadiSys
Drives
  Chinon America
  Hitachi America Ltd.
  Iomega
  Laser Magnetic Storage
  Matsushita Electric
  Maxoptix
  Maxtor
  Micropolis Corporation
  Mitsubishi Electronics America
  Panasonic Communications &
    Systems Co.
  Plextor
  Quantum Corporation
  Seagate Technology
  Sony Electronics
  SyQuest Technology, Inc.
  Teac America, Inc.
  Toray
  Western Digital Corporation

Flash (nondisk) Storage
  SanDisk Corporation
Hardware
  Amecon, Inc.
  Antec, Inc.
Servers
  Amdahl
  Auspex Systems, Inc.
  Berkeley Software Design
  Sequent Computer Systems, Inc.
  Tandem Computers
  Tangent Computers
Single Board
  HiTech Equipment Corporation
Supercomputers
  Cray Research, Inc.
  Tera Computer Company
Systems
  Acer America Corporation
  Advanced Logic Research (ALR)
  Alps Electric
  Apple Computer
  Aris Microsystems
  Aspen Systems, Inc.
  AST Research
  Austin Computers
  Avnet
  Compaq Computer Corporation
  Computer Discount Warehouse
  Convex
  CyberMax Computer, Inc.
  DAKCO
  Data General
  Daystar Digital
  Dell Computer Corporation
  Digital Equipment Corporation
  Direct Wave
  DTI Factory Outlet
  Encore Computer Corporation
  Envision Interactive Computers
  EPS Technologies, Inc.
  Everex Systems

Express Systems
Fujitsu America, Inc.
Galaxy Computers, Inc.
Gateway 2000
HDSS Computer
Hewlett-Packard Company
Hyundai Electronics America
IBM
Jem Computers
Leapfrog Lab, Inc.
Maximus Computers, Inc.
Micron Electronics, Inc.
Midwest Micro
MPC Technologies, Inc.
National MicroComputers
Northgate Computer Systems, Inc.
Occo Enterprises, Inc.
Industrial
APPRO International, Inc.
Interlogic
Ziatech Corporation
Modems
Boca Research
Cardinal Technologies
Gammalink
Hayes Microcomputer Products
Megahertz Corporation
Practical Peripherals, Inc.
Supra Communications
Telebit
Telebyte Technology, Inc.
U.S. Robotics (USR)
Zoom Telephonics
ZyXEL
Motherboards
Alaris
American Megatrends, Inc. (AMI)
Hauppauge Computer Works, Inc.
Micronics Computers

Multimedia
Amecon, Inc.
Diamond Multimedia
IMSI
Infusion Systems, Ltd.
Jazz Multimedia
Silicom Electronics, Inc.
Sony Electronics
Power Supplies
See Equipment/Power Supplies.
Printers
Apple Computer
Brother International Corporation
Canon USA
Citizen America Corporation
C. Itoh
Dataproducts, Inc.
Epson America, Inc.
Fargo Electronics
Hewlett-Packard Company
Lexmark International
Okidata
Olivetti
Pacific Data Products
Panasonic Communications & Systems Co.
QMS
Ricoh Company, Ltd.
Star Micronics
Tektronix
Refurbished
MicroExchange
Tredex
Odyssey Computing
Packard Bell
PCs Compleat
Peak Computing
PerComp Microsystems, Inc.
Quantex Microsystems, Inc.
Quay Computers
Samsung Electronics America

Shop Direct, Inc.
Swan Technologies Corporation
Tagram Systems Corporation
Tandy Corporatrion
TexasMicro
Toshiba America
UMAX Technologies
Unisys Corporation
Vektron International
Xerox
Zenith Data Systems
Zenon Technology, Inc.
Zeos International, Ltd.
Telephony
Talking Technologies
VME
Heurikon Corporation
Workstations
HaL Computer Systems
HDS Network Systems, Inc.
RDI Computer
Silicon Graphics
Sun Microsystems

CONTROLLERS

Drive
Adaptec
AdvanSys
Arco Electronics, Inc.
DTC
Promise Technology
PSI-Squared, Ltd.
Quadram
Realtek
Silicon Valley Research
Tecmar, Inc.
Tekram Technology
Ultrastor Corporation
I/O
Central Data

Network
Adaptec
Interphase Corporation
IQ Technologies, Inc.
SCSI
Advanced Storage Concepts
BusLogic
Distributed Processing Technology
(DPT)
Micro Design International
Mylex
Qlogic Corporation
Rancho Technology, Inc.
Ultrastor Corporation
Video
ATI Technologies
Boca Research
Cardinal Technologies
Compression Labs, Inc.
Creative Labs
Diamond Multimedia
Hauppauge Computer Works,
Inc.
Hercules Computer Technology
Matrox
Media Vision
Number Nine Visual Technology
Orchid Technology
Paradise
STB Systems
Tekram Technology
Tseng Labs, Inc.

DIAGNOSTICS

Benchmarks
Bapco
Hardware
AllMicro
Data Depot, Inc.
Micro2000

Telecommunications Techniques
  Corp.
Teradyne
Software
  AllMicro
  American Megatrends, Inc.
    (AMI)
  AnaTek Corporation
  Central Point Software, Inc.
  Data Depot, Inc.
  Micro2000
  MicroSystems Development
  Ontrack Computer Systems, Inc.
  PowerQuest
  Reynolds Data Recovery
  Total Recall
  WaterGate Software
  Wave Tech
Virus
  McAfee and Associates
  Symantec Corporation
  TouchStone Software Corporation

DISTRIBUTOR

Components
  Allied Electronics
  Alternate Source Components
  Bell Industries
  Circuit Specialists, Inc.
  Component Distributors, Inc.
  Daybreak Communications
  Digi-Key Corporation
  Ecliptek
  Electronics Marketing Corpora-
    tion
  Electronix Corporation
  Farnell Components
  FG Commodity Electronics
  Future Electronics
  Halfin

Hamilton-Hallmark
Hawk Electronics
Hi-Tech Component Distributors,
  Inc.
Hi-Tech Surplus
Insight Electronics
Jaco Electronics
Jameco Computer Products
JDR Microdevices
Marsh Electronics
Marshall Industries
Mitronics
Mouser Electronics
National Parts Depot
Nu Horizons Electronics
PartNET
Peerless Radio Corporation
Pioneer Electronics
Powell Electronics, Inc.
Reptron Electronics
RicheyCypress Electronics
Sager Electronics
SEI
Summit Distributors
Techlock Distributing
Time Electronics
Triangle Electronics
Webb Distribution
Drives
  APS Technologies
  Core International
  Corporate Systems Center
  DC Drives
  Exabyte Corporation
  Hard Drives Northwest
  Micro Solutions
  Storage USA
Electro-Mechanical Components
  Sterling Instruments
Hobby
  PAiA Electronics

Memory
  ACIS, Inc.
  Kingston Technology
  Mission Electronics
  Simple Technology
  Simtek Corporation
  TechWorks
  TEM Computers
  Upgrades Unlimited
  Visiontek
  Windows Memory Corporation
Motherboards
  *American Micro Products Tech-
    nology*
  First Computer Systems
  Motherboard Discount Center
PC Assemblies (i.e. drives, adapters,
  video boards)
  A2Z Computers
  *American Micro Solutions
    (AMS)*
  CDW Computer Centers, Inc.
  CMO
  Comp-U-Plus
  CompUSA Direct
  ComputAbility
  Computer Marketplace, Inc.
  Data Comm Warehouse
  *Data Technology Corporation
    (DTC)*
  Elek-Tek, Inc.
  Hartford Computer Group
  Ingram Micro
  Insight Direct
  MegaHaus
  Merisel
  MicroXperts
  MiniMicro
  *Multiple Brand Superstore
    (MBS)*
  Omni Data Communications
  PC Universe
  PC Warehouse

  Peripherals Unlimited
  Quanta Micro Corporation
  SBI Computer Products
  Spectrum Trading
  Sunshine Computers
  Surplus Software, Inc.
  TC Computers
  Team American
  Tech Data Corporation
  Terabyte
  Tyan Computer
  US Computer Direct Corporation
Scanners
  Envisions
  Imacon
  Visioneer
SIMM Adapters
  Sermax Corporation
  SIMM Saver Technology, Inc.
  Tucson Computer Corporation
Software
  Data Comm Warehouse
  Micro Warehouse
  Provantage Corporation
  Public Software Library (PsL)
  Surplus Software, Inc.
Supplies
  Computer Gate International

EQUIPMENT

Cable
  Advent Electronics
  Applied Innovation
  General Instrument Corporation
  Texscan Corporation
Communication
  Copper Electronics
  Cybex Corporation
  Dialogic
  Ericsson
  Motorola
  Radio Control Systems, Inc.

Tadiran
Tone Commander
Unicon, Inc.
Vive Synergies, Inc.
Dongles
Rainbow Technologies
EMI
Schaffner EMC
Emulation
Embedded Support Tools Corp.
Huntsville Microsystems, Inc.
Lauterbach Datentechnik
Softaid, Inc.
Virtual Machine Works
Encryption
Marx International, Inc.
Imaging
Alden Electronics
ComCom Systems
ISG Technologies
Picker International, Inc.
Precision Digital Images
SoftNet Systems
Voxel
Xionics
Industrial
Allen-Bradley
Dove Systems
Lamp Technology, Inc.
Z-World Engineering
Instruments
Ferran Scientific
Kaman Instrumentation Corpo-
ration
Keithley Instruments
National Instruments
TSI
Varian
Laser Scanning
Symbol Technologies
Magnetic Stripe
Mag-Tek
Power Products

American Power Conversion
(APC)
Argus Technologies, Ltd.
Astec
Electrotek Concepts, Inc.
General Electric
Panamax
Power Convertibles
Power Trends
Vicor Corporation
Power Supplies
Advance International
Aslan Computer
Phihong
PLC (Programmable Logic Con-
troller)
AEG Schneider Automation
Printer Sharing
Raritan Computer, Inc.
Programming
Aaeon
ITU Technologies
Needham's Electronics
Prototyping
Vector Electronics Company
Satellite
Advent Electronics
Security
D&S Technologies
Test
Aaeon
Avtech Electrosystems, Ltd.
CST
Extron
Fluke
GlobeTech International
I-Tech Corporation
JECH Tech
LeCroy Corporation
Weather
Alden Electronics
Video
AVerMedia

Connectix Corporation
Digital Vision, Inc.
Display Tech Multimedia
General Instrument Corporation
miro Computer Products
VideoLabs
Videonics

ICs

Analog
Analog Devices
Brooktree Corporation
Burr-Brown
Exar Corporation
International Rectifier
Linear Technology
Maxim
National Semiconductor
BIOS
American Megatrends, Inc.
(AMI)
Asus
Award Software International
DTK Computer, Inc.
MicroFirmware
(Mr. BIOS) Microid Research,
Inc.
Phoenix Technologies, Ltd.
TTi Technologies, Inc.
Unicore Software, Inc.
Communication
Level One Communication
Mitel Semiconductor
PMC-Sierra
Rockwell International
Seeq Technology
Standard Microsystems Corpora-
tion
Teltone
TransSwitch
Winbond Electronics Corporation

Custom Design
Advanced Microelectronics
DSP
Graychip
Logic
Actel Corporation
Altera Corporation
AMCC
Amtel Corporation
Avance Logic
Chips & Technologies
Chromatic Research
Chrontel
Cirrus Logic
Cypress Semiconductor
Dallas Semiconductor
Elantec
Fujitsu America, Inc.
Hitachi America Ltd.
IMP
Information Storage Devices
Integrated Device Technology
Lattice Semiconductor
LSI Logic Corporation
Micron Technology
Music Semiconductors
NCR Microelectronics
NVidia Corporation
OPTI, Inc.
Quality Semiconductor
Rambus
Ramtron International
Texas Instruments
V3 Semiconductor
WSI, Inc.
Xicor
Xilinx
Zilog
Microprocessors
Advanced Micro Devices (AMD)
Cyrix
IBM

Intel Corporation
Motorola
NexGen, Inc.
Weitek Corporation
Mixed-Signal
Harris Semiconductor
Integrated Telecom Technology
MX.COM, Inc.
Silicon Systems, Inc.
Synergy Semiconductor
Power Management
Benchmarq Microelectronics
RISC Processors
Advanced RISC Machines
Microchip Technology
MIPS Technologies
SPARC Processors
Ross Technology, Inc.

MANUFACTURER

Components
Apem
Aromat Corporation
Avex Electronics, Inc.
California Switch & Signal
Cermetek Microelectronics
Corcom
CTS
Likom
Marmah Magnetics, Inc.
Methode Electronics
Microsemi
Microwave Filter Co., Inc.
Microwave Power, Inc.
Molecular OptoElectronics Corporation
Oki America, Inc.
Phillips ECG
Plessey Semiconductors
Republic Electronics
SaRonix

SB Electronics, Inc.
Schurter
Shallco, Inc.
TEMIC (Siliconix)
Thomas & Betts
Trilithic
TRW
Visual Communications Company
Wenzel Associates
Custom
Atec, Inc.
Bromley Instruments
EJE Research
EMD Associates
FlashPak
Hunter Technology Corporation
Insulectro
Maxwell Laboratories, Inc.
Micro Technology Services
Pacific Microelectronics
TLG Electronics
ICs
Chip Express
Delco Electronics
Direct Silicon Design
Integrated Circuit Design Concepts
Lansdale Semiconductor
Quickturn
SGS-Thompson Microelectronics
Siemens USA
Silicon Engineering
S-MOS Systems
Materials
Corning
CSI (Colorado Superconductor Inc.)
Southwire
Teldor
PCB
Advanced Quick Circuits
Alberta Printed Circuits, Ltd.

Circuit Works
Collins Printed Circuits
Milplex Circuits, Inc.
Miltec Electronics, Inc.
Nu-Tronic Circuit Company
Standard Printed Circuits, Inc.
U.S. Technologies, Inc.
Sensors
HyCalNet
Keekor USA
Lundahl Instruments
Sensor Development, Inc.
Software
Corporate Disk
Tape Heads
Magnetic Technology

NETWORKING

Cabling
Hybrid Networks
CD-ROM
Axis Communications
Enterprise
ADC Telecommunications
Madge Networks
Hardware
3Com Corporation
*ACC (Advanced Computer Com-
munication)*
*ACC (Associated Computer Con-
sultants)*
Accton
Agile Networks
Ainsworth Technology
Alantec
Amber Wave Systems
Asante Technologies, Inc.
Ascend Communications
Bay Networks
Bellcore
Black Box Corporation

Cisco
Cogent Data
*Computer Network Technology
Corp.*
CrossComm Corporation
Develcon
DigiBoard
D-Link Systems, Inc.
FastComm Communications
Fibronics
General DataComm
Imatek, Inc.
IMC
ISDN Systems Corporation
Kalpana
Klever
LANSource
Larscom
MICOM
Microcom
MPR Teltech
NetEdge
Netrix Corporation
Network Computing Devices
Newbridge
Peripheral Technology Group, Inc.
Plaintree Systems
Proteon
Racal-Datacom Corporation
SBE, Inc.
Shiva Corporation
Skyline Technology
*Standard Microsystems Corpora-
tion*
SVEC Computer Corporation
Telenex Corporation
Thinking Machines Corporation
UB Networks
Wall Data, Inc.
Western Micro Technology
Xircom
Xylogics, Inc.

Printing
  Axis Communications
Software
  Asante Technologies, Inc.
  Banyan Systems
  Cabletron Systems, Inc.
  Cheyenne Software
  Cogent Data
  Connectware, Inc.
  Farallon Computing
  Forte
  Funk Software, Inc.
  Gupta
  Integrated Systems, Inc.
  Microcom
  Network Computing Devices
  Novell, Inc.
  Performance Technology
  *Peripheral Technology Group,
    Inc.*
  Persoft, Inc.
  Shiva Corporation
  Sybase
  Thinking Machines Corporation
  Wall Data, Inc.
  Western Micro Technology

PERIPHERALS

Data Acquisition
  ADAC
  Adept Scientific
  Intelligent Instrumentation
  Silicon Soft
Data Storage
  3M Data Storage Products
DSP
  Ariel Corporation
Input Devices
  Key Tronic Corporation
  NMB Technologies
  Tanisys Technology

  Texas Industrial Peripherals
LCDs
  Okaya Electric America, Inc.
  Sharp Electronics
Memory
  ALL Computer
  Autotime Corporation
Monitors
  Amdek Corporation
  Hyundai Electronics America
  Magnavox
  Nanao USA
  NEC Technologies, Inc.
  Nokia Display Products
  Portrait Display Labs
  ViewSonic Corporation
  Wen Technology
  Wyse Technology
RISC Workstations
  Abstract Technologies
Sound
  Adcom
  Atlas Soundolier
  Creative Labs
  Dolby Laboratories
  Doxa Audio Development
  DSP Group
  Eastern Acoustic Works
  Fisher
  Ivie Technologies
  Marantz
  Milbert Amplifiers, Inc.
  Q-Sound
  Spatializer Audio Labs
  SRS Labs
  Sumo
  Sweetwater Sound, Inc.
Sound Boards
  Aztech Labs, Inc.
  Creative Labs
  Ensoniq
  Turtle Beach Systems

Speakers
Altec Lansing Corporation
Definitive Technology
TOA Electronics, Inc.
Supercomputers
Alta Technology
Tape
*Advanced Digital Information
    Corp.*
Alloy Computer Products
Colorado Memory Systems, Inc.
Columbia Data Products
Giga Trend, Inc.
Lots Technology, Inc.
Maynard Electronics, Inc.
Mountain Network Solutions, Inc.
Overland Data
TAPEDISK Corporation

PUBLISHER

Computer
Addison-Wesley Publishing
Annabooks
Cera Research
Computer Connections
Computer Currents
Computer Shopper
Disaster Recovery Journal (DRJ)
Imaging Magazine
Macmillan Computer Publishing
McGraw-Hill
PC Catalog
PC Computing
Pinnacle Publishing, Inc.
The Computer Journal
Ventana Communications Group
Windows Magazine
Computer Repair
Dynamic Learning Systems
Electronics
Butterworth Heinemann

Circuit Cellar, Inc.
EE Times
McGraw-Hill
Processor Magazine
Internet
National Computer Tectonics
Peachpit Press
Multimedia
Betacorp Multimedia
Software
Walnut Creek CD-ROM

RECYCLING

Fill Kits
Computer Friends
Toner Cartridge
Environmental Laser

SECURITY

Hardware
Ademco
Stone Technologies

SERVICES

Anti-Piracy
*Software Publisher's Association
    (SPA)*
Data Conversion
Pivar Computing Services, Inc.
Data Recovery
Columbia Data Products
Ontrack Computer Systems, Inc.
Total Recall
Education
CAST
Data-Tech Institute
ICS Learning Systems
Keystone Learning Systems Corp.
TyLink Corporation

Information
  Nanothinc
  Sparc International
On-line
  America Online
  CompuServe
  Computer Mail Services
  Global Village Communication
  MecklerWeb
  MegaSoft
  NetCom
  Prodigy
  SPRY
  UUNet Technologies
Repair
  Continental Resources
  Electroservice Laboratories
  Micro-Vision
  Scope Systems
  Sprague Magnetics, Inc.
Research
  Research Triangle Institute
  SAIC
  SystemWare
Standards
  PCMCIA

SIMULATION

Circuit CAD
  Advanced Microcomputer Systems
  Aldec, Inc.
  Ansoft Corporation
  Antares Corporation
  Cadence
  Capilano Computing Systems
  Cascade
  Chrysalis Symbolic Design, Inc.
  Data I/O
  dTb Software
  Eagle Design Automation
  ELANIX

Frontline Design Automation
Interactive Image Technologies
  Ltd.
Intergraph
Intusoft
Ivex Design International
Meta-Software
MicroSim Corporation
Model Technology
Optotek
OrCad
Protel Technology
Synopsys
TurboSim (Island Logix, Inc.)
VeriBest, Inc.
VIEWLogic
Visionics
Web Laboratories
IC Models
  Library Technologies, Inc.
  Topdown
Training
  Hantro

SOFTWARE UTILITIES

Business
  Allegro New Media
  Blue Sky Software Corporation
  B-Plan
  Claris Corporation
  Computer Associates
  ConnectSoft
  Delorme
  Deneba Software
  Ditek Software Corporation
  Folio Corporation
  HyperDesk Corporation
  Janna Systems, Inc.
  KnowledgeBroker, Inc.
  Macromedia
  Mastersoft, Inc.

MECA Software
Micro Logic
Napersoft, Inc.
PC-Kwik Corporation
Pipeline Network
Plannet Crafters
Radius
ServiceWare Corporation
Software Publishing Corporation
   (SPC)
Visio Corporation
Word Perfect
CAD
   Autodesk
   Caligari
   Mentor Graphics
   Micrografx
   Numera Software
CASE
   Iconix
Communication
   Datastorm Technologies
   Delrina Corporation
   ELAN Software Corporation
   Excalibur Communications
   FTP Software
   Galacticomm
   Hilgraeve
   Metz Software
   MobileWare
   Mustang Software
   Qualcomm
   Traveling Software
Databases
   askSam Systems
   Borland International
   Lotus Development Corporation
   Oracle
Data Compression
   PKWARE, Inc.
   Stac Electronics
   WinZip

Development
   20/20 Corporation
   Aladdin Knowledge Systems,
      Ltd.
   Applied Microsystems Corporation
   Borland International
   Bristol Technology
   Capsoft Development Corporation
   Criterion Software
   DSPnet
   Intec Inoventures, Inc.
   MicroEdge
   Micro Focus, Ltd.
   Microtek International
   Mortice Kern Systems (MKS)
   Nu-Mega Technologies, Inc.
   ObjecTime Limited
   Orion Instruments
   Parallax
   Powersoft
   QNX Software Systems, Ltd.
   Spectron
   Stylus Product Group
   Taligent
   TechTools, Inc.
   TeraTech
   VenturCom, Inc.
   Visionware, Ltd.
   Watcom International
   Wind River Systems
DSP
   Hyperception
DTP
   Adobe Systems
   Corel Corporation
   Fractal Design Corporation
   Frame Technology Corporation
   Harlequin
   Inset Systems
Embedded Product Development
   Accelerated Technology, Inc.
   Micro Digital, Inc.

U.S. Software
Verilog, Inc.
Encryption
Aliroo, Ltd.
Fonts
Bitstream, Inc.
Imaging
Adobe Systems
Caere Corporation
Elastic Reality
HSC Software
InfoImaging Technologies, Inc.
LEAD Technologies
Zylab
Internet
American Cybernetics
Attachmate
Beame and Whiteside Software
(Hummingbird Communications Ltd.)
CE Software
Connect, Inc.
Enterprise Integration Technologies
Farallon Computing
Frontier Technologies Corporation
Iconovex
Insignia Solutions
Intercon
Mediatrix Peripherals
Moon Valley Software
Morning Star Technologies
Nesbit Software
NetManage, Inc.
Netscape Communications Corporation
NeXT Computer
Open Market
Quarterdeck Corporation
Spyglass
Starfish

Memory Management
Qualitas, Inc.
Quarterdeck Corporation
Operating Systems (OS)
IBM
Industrial Programming, Inc.
JMI Software Systems, Inc.
Microsoft Corporation
Printer
Extended Systems
Reference
Allegro New Media
Micro House International
Softinfo
Torch USA
Scientific
MathSoft
Mathsource
MathWare
MathWorks
MicroMath Scientific Software
Video Compression
Motion Pixels
Voice Recognition
Verbex Voice Systems

THERMAL PRODUCTS

Heat Sinks
Aavid Thermal Technologies, Inc.
FTS Systems
Tellurex
Software
K&K Associates

TOOLS

Data Compression
Advanced Hardware Architectures
InfoChip Systems, Inc.

Error Correction
Advanced Hardware Architectures

VIDEO

Controllers
See *Controllers/Video*.
Multimedia Equipment
Abekas Video Systems
PC-TV Conversion
Advanced Digital Systems

On-Line Glossary

The world of the Internet is laden with specialized acronyms and networking terms. If you work on-line for any length of time, you're probably going to need a glossary, so here is a list of popular terms.

10BaseT A variant of Ethernet that allows stations to be attached via twisted pair cable. See Ethernet, twisted pair.

Acceptable Use Policy (AUP) Many transit networks have policies that restrict the use to which the network can be put. A well known example is NSFNET's AUP, which does not allow commercial use. Enforcement of AUPs varies with the network.

Access Control List (ACL) Most network security systems operate by allowing selective use of services. An Access Control List is the usual means by which access to, and denial of, services is controlled. It is simply a list of the services available, each with a list of the hosts permitted to use the service.

acknowledgment (ACK) A type of message sent to indicate that a block of data arrived at its destination without error.

address There are three types of addresses in common use within the Internet. They are e-mail address, IP, Internet address, and hardware or MAC address.

address mask A bit mask used to identify which bits in an IP address correspond to the network and subnet portions of the address. This mask is often referred to as the *subnet mask* because the network portion of the address can be determined by the encoding inherent in an IP address.

address resolution Conversion of an Internet address into the corresponding physical address.

Address Resolution Protocol (ARP) Used to dynamically discover the low-level physical network hardware address that corresponds to the high-level IP address for a given host. ARP is limited to physical network systems that support broadcast packets that can be heard by all hosts on the network.

Advanced Research Projects Agency Network (ARPANET) A pioneering long-haul network funded by ARPA (now DARPA). It served as the basis for early networking research, as well as a central backbone during the development of the Internet. The ARPANET consisted of individual packet switching computers interconnected by leased lines.

agent In the client-server model, the part of the system that performs information preparation and exchange on behalf of a client or server application.

alias A name, usually short and easy to remember, that is translated into another name, usually long and difficult to remember.

American National Standards Institute (ANSI) This organization is responsible for approving U.S. standards in many areas, including computers and communications. Standards approved by this organization are often called ANSI standards (e.g., ANSI C is the version of the C language approved by ANSI). ANSI is a member of ISO.

American Standard Code for Information Interchange (ASCII) A standard character-to-number encoding widely used in the computer industry.

anonymous FTP Anonymous FTP allows a user to retrieve documents, files, programs, and other archived data from anywhere in the Internet without having to establish a user ID and password. By using the special user ID of "anonymous" the network user will bypass local security checks and will have access to publicly accessible files on the remote system.

application A program that performs a function directly for a user. FTP, mail, and Telnet clients are examples of network applications.

Application Program Interface (API) A set of calling conventions that define how a service is invoked through a software package.

Archie   A system to automatically gather, index, and serve information on the Internet. The initial implementation of Archie provided an indexed directory of filenames from all anonymous FTP archives on the Internet. Later versions provide other collections of information.

archive site   A machine that provides access to a collection of files across the Internet. An "anonymous FTP archive site," for example, provides access to this material via the FTP protocol.

Asynchronous Transfer Mode (ATM)   A method for the dynamic allocation of bandwidth using a fixed-size packet (called a *cell*). ATM is also known as *fast packet*.

backbone   The top level in a hierarchical network. Stub and transit networks that connect to the same backbone are guaranteed to be interconnected.

bandwidth   The difference, in Hertz (Hz), between the highest and lowest frequencies of a transmission channel. Also meant to represent the amount of data that can be sent through a given communications circuit.

bang path   A series of machine names used to direct electronic mail from one user to another, typically by specifying an explicit UUCP path through which the mail is to be routed.

BCNU   A typical e-mail acronym for "Be Seein' You."

Berkeley Internet Name Domain (BIND)   Implementation of a DNS server developed and distributed by the University of California at Berkeley. Many Internet hosts run BIND, and it is the ancestor of many commercial BIND implementations.

Birds Of a Feather (BOF)   A Birds Of a Feather (flocking together) is an informal discussion group. It is formed ad hoc to consider a specific issue, and therefore, has a narrow focus.

Bitnet   An academic computer network that provides interactive electronic mail and file-transfer services, using a store-and-forward protocol, based on IBM Network Job Entry protocols. Bitnet-II encapsulates the Bitnet protocol within IP packets and depends on the Internet to route them.

bounce   The return of a piece of mail because of an error in its delivery.

bridge   A device that forwards traffic between network segments based on data link layer information. These segments would have a common network layer address.

BTW   A typical e-mail acronym for "By The Way."

Bulletin Board System (BBS)   A computer and associated software that typically provides electronic messaging services, archives of files, and any other services or activities of interest to the bulletin board system's operator. Although BBSs have traditionally been the domain of hobbyists, an increasing number of BBSs are connected directly to the Internet, and many BBSs are currently operated by government, educational, and research institutions.

Campus Wide Information System (CWIS) A CWIS makes information and services publicly available on campus via kiosks, and makes interactive computing available via kiosks, interactive computing systems, and campus networks. Services routinely include directory information, calendars, bulletin boards, and databases.

checksum A computed value that is dependent upon the contents of a packet. This value is sent along with the packet when it is transmitted. The receiving system computes a new checksum based on the received data and compares this value with the one sent with the packet. If the two values are the same, the receiver has a high degree of confidence that the data was received correctly.

client A computer system or process that requests a service of another computer system or process. A workstation requesting the contents of a file from a file server is a client of the file server.

client-server model A common way to describe the paradigm of many network protocols. Examples include the name-server/name-resolver relationship in DNS, and the file-server/file-client relationship in NFS.

Coalition for Networked Information (CNI) A consortium formed by American Research Libraries, CAUSE, and EDUCOM to promote the creation of, and access to, information resources in networked environments in order to enrich scholarship and enhance intellectual productivity.

Comite Consultatif International de Telegraphique et Telephonique (CCITT) This organization is part of the International Telecommunications Union (ITU) and is responsible for making technical recommendations about telephone and data communications systems. Every four years CCITT holds plenary sessions where they adopt new standards; the most recent was in 1992.

Computer Emergency Response Team (CERT) The CERT was formed by DARPA in November 1988 in response to the needs exhibited during the Internet worm incident. The CERT charter is to work with the Internet community to facilitate its response to computer security events involving Internet hosts, to take proactive steps to raise the community's awareness of computer security issues, and to conduct research targeted at improving the security of existing systems. CERT products and services include 24-hour technical assistance for responding to computer security incidents, product vulnerability assistance, technical documents, and tutorials. In addition, the team maintains a number of mailing lists (including one for CERT Advisories), and provides an anonymous FTP server, at cert.org, where security-related documents and tools are archived. The CERT can be reached by email at cert@cert.org and by telephone at 412-268-7090 (24-hour hotline).

congestion A condition that occurs when the offered load exceeds the capacity of a data communication path.

connection-oriented   The data communication method in which communication proceeds through three well-defined phases: connection establishment, data transfer, and connection release. TCP is a connection-oriented protocol.

connectionless   The data communication method in which communication occurs between hosts with no previous setup. Packets between two hosts might take different routes, as each is independent of the other. UDP is a connectionless protocol.

Coordinating Committee for Intercontinental Research Networks (CCIRN)   A committee that includes the United States FNC and its counterparts in North America and Europe. Co-chaired by the executive directors of the FNC and the European Association of Research Networks (RARE), the CCIRN provides a forum for cooperative planning among the principal North American and European research networking bodies.

Corporation for Research and Educational Networking (CREN)   This organization was formed in October 1989, when Bitnet and CSNET (Computer and Science NETwork) were combined under one administrative authority. CSNET is no longer operational, but CREN still runs Bitnet.

cracker   An individual who attempts to access computer systems without authorization. These individuals are often malicious, as opposed to hackers, and have many means at their disposal for breaking into a system.

cyberspace   A term coined by William Gibson in his fantasy novel Neuromancer to describe the "world" of computers, and the society that gathers around them.

Cyclic Redundancy Check (CRC)   A number derived from a set of data that will be transmitted. By recalculating the CRC at the remote end and comparing it to the value originally transmitted, the receiving node can detect some types of transmission errors.

Data Encryption Key (DEK)   Used for the encryption of message text and for the computation of message integrity checks (signatures).

Data Encryption Standard (DES)   A popular, standard encryption scheme, which is illegal to export out of the US.

DECnet   A proprietary network protocol designed by Digital Equipment Corporation. The functionality of each phase of the implementation, such as Phase IV and Phase V, is different.

Defense Advanced Research Projects Agency (DARPA)   An agency of the U.S. Department of Defense responsible for the development of new technology for use by the military. DARPA (formerly known as ARPA) was responsible for funding much of the development of the Internet we know today, including the Berkeley version of Unix and TCP/IP.

Defense Data Network (DDN)   A global communications network serving the US Department of Defense composed of MILNET, other portions of the Internet, and classified networks that are not part of the Internet. The DDN is used to connect military installations and is managed by the Defense Information Systems Agency.

Defense Information Systems Agency (DISA) Formerly called the Defense Communications Agency (DCA), this is the government agency responsible for managing the DDN portion of the Internet, including the MILNET. Currently, DISA administers the DDN, and supports the user-assistance services of the DDN NIC.

dialup A temporary (as opposed to dedicated) connection between machines established over a standard phone line. BBS and individual Internet access is typically dialup in nature.

Distributed Computing Environment (DCE) An architecture of standard programming interfaces, conventions, and server functionality (i.e., naming, distributed file system, or remote procedure call) for distributing applications transparently across networks of heterogeneous computers. Promoted and controlled by the Open Software Foundation (OSF), a consortium led by Digital, IBM, and Hewlett-Packard.

Domain Name System (DNS) The DNS is a general-purpose distributed, replicated, data query service. The principal use is the lookup of host IP addresses based on host names. The style of host names now used in the Internet is called *domain name,* because they are the style of names used to look up anything in the DNS. Some important domains are: .COM (commercial), .EDU (educational), .NET (network operations), .GOV (U.S. government), and .MIL (U.S. military). Most countries also have a domain. For example, .US (United States), .UK (United Kingdom), or .AU (Australia).

dot address (dotted decimal notation) Refers to the common notation for IP addresses of the form A.B.C.D; each letter represents, in decimal, one byte of a four-byte IP address.

Electronic Frontier Foundation (EFF) A foundation established to address social and legal issues arising from the impact on society of the increasingly pervasive use of computers as a means of communication and information distribution.

electronic mail (e-mail) A system whereby a computer user can exchange messages with other computer users (or groups of users) via a communications network. Electronic mail is one of the most popular uses of the Internet.

e-mail address The domain-based or UUCP address that is used to send electronic mail to a specified destination. For example my address is sbigelow@cerfnet.com.

encryption The manipulation of a packet's data in order to prevent any but the intended recipient from reading that data. There are many types of data encryption, and they are the basis of network security.

Ethernet   A 10-Mb/s standard for LANs, initially developed by Xerox and later refined by Digital, Intel, and Xerox (DIX). All hosts are connected to a coaxial cable where they contend for network access using a Carrier Sense Multiple Access with Collision Detection (CSMA/CD) paradigm.

European Academic and Research Network (EARN)   A network connecting European academic and research institutions with electronic mail and file-transfer services using the Bitnet protocol.

Extended Binary Coded Decimal Interchange Code (EBCDIC)   A standard character-to-number encoding used primarily by IBM computer systems.

FARNET   A nonprofit corporation, established in 1987, whose mission is to advance the use of computer networks to improve research and education.

FAQ   A frequently asked question.

Fiber Distributed Data Interface (FDDI)   A high-speed (100Mb/s) LAN standard. The underlying medium is fiber optics, and the topology is a dual-attached, counter-rotating token ring.

file transfer   The copying of a file from one computer to another over a computer network.

File Transfer Protocol (FTP)   A protocol that allows a user on one host to access and transfer files to and from another host over a network. Also, FTP is usually the name of the program the user invokes to execute the protocol.

finger   A program that displays information about a particular user, or all users, logged on the local system or on a remote system. It typically shows full name, last login time, idle time, terminal line, and terminal location (where applicable). It also might display plan and project files left by the user.

flame   A strong opinion and/or criticism of something, usually as a frank inflammatory statement, in an electronic mail message. It is common to precede a flame with an indication of pending fire (i.e., FLAME ON!). Flame wars occur when people start flaming other people for flaming when they shouldn't have.

freenet   Community-based bulletin board system with email, information services, interactive communications, and conferencing. Freenets are funded and operated by individuals and volunteers, in one sense they are like public television. They are part of the National Public Telecomputing Network (NPTN), an organization based in Cleveland, Ohio, devoted to making computer telecommunication and networking services as freely available as public libraries.

Fully Qualified Domain Name (FQDN)   The FQDN is the full name of a system, rather than just its host name. For example, "venera" is a host name and "venera.isi.edu" is an FQDN.

gateway   The term *router* is now used in place of the original definition of "gateway." Currently, a gateway is a communications device/program that passes data between networks having similar functions but dissimilar implementations. This should not be confused with a protocol converter. By this definition, a router is a layer 3 (network layer) gateway, and a mail gateway is a layer 7 (application layer) gateway.

Gopher A distributed information service that makes available hierarchical collections of information across the Internet. Gopher uses a simple protocol that allows a single Gopher client to access information from any accessible Gopher server, providing the user with a single "Gopher space" of information. Public domain versions of the client and server are available.

hacker A person who delights in having an intimate understanding of the internal workings of a system, computers, and computer networks in particular. The term is often misused in a negative context, where *cracker* would be the correct term.

hierarchical routing The complex problem of routing on large networks can be simplified by reducing the size of the networks. This is accomplished by breaking a network into a hierarchy of networks, where each level is responsible for its own routing. The Internet has basically three levels: the backbones, the mid-levels, and the stub networks. The backbones know how to route between the mid-levels, the mid-levels know how to route between the sites, and each site (being an autonomous system) knows how to route internally.

High Performance Computing and Communications (HPCC) High-performance computing encompasses advanced computing, communications, and information technologies, including scientific workstations, supercomputer systems, high-speed networks, special-purpose and experimental systems, the new generation of large-scale parallel systems, and application and systems software with all components well integrated and linked over a high-speed network.

High-Performance Parallel Interface (HIPPI) An emerging ANSI standard that extends the computer bus over fairly short distances at speeds of 800 and 1600 Mb/s. HIPPI is often used in a computer room to connect a supercomputer to routers, frame buffers, mass-storage peripherals, and other computers.

host A computer that allows users to communicate with other host computers on a network. Individual users communicate by using application programs, such as electronic mail, Telnet, and FTP.

hub A device connected to several other devices. In ARCnet, a hub is used to connect several computers together. In a message handling service, a hub is used for the transfer of messages across the network.

IMHO An e-mail acronym for "In My Humble Opinion."

Integrated Services Digital Network (ISDN) An emerging technology that is beginning to be offered by the telephone carriers of the world.

ISDN combines voice and digital network services in a single medium, making it possible to offer customers digital data services as well as voice connections through a single "wire." The standards that define ISDN are specified by CCITT.

Interagency Interim National Research and Education Network (IINREN) An evolving operating network system. Near-term research and development activities will provide for the smooth evolution of this networking infrastructure into the future gigabit NREN.

International Organization for Standardization (ISO) A voluntary, nontreaty organization founded in 1946 that is responsible for creating international standards in many areas, including computers and communications. Its members are the national standards organizations of the 89 member countries, including ANSI for the U.S.

Internet address An IP address that uniquely identifies a node on the Internet.

Internet Architecture Board (IAB) The technical body that oversees the development of the Internet suite of protocols. It has two task forces: the IETF and the IRTF. "IAB" previously stood for Internet Activities Board.

Internet Control Message Protocol (ICMP) An extension to the Internet Protocol. It allows for the generation of error messages, test packets, and informational messages related to IP.

Internet Engineering Steering Group (IESG) The IESG is composed of the IETF Area Directors and the IETF Chair. It provides the first technical review of Internet standards and is responsible for day-to-day "management" of the IETF.

Internet Engineering Task Force (IETF) A large, open community of network designers, operators, vendors, and researchers whose purpose is to coordinate the operation, management, and evolution of the Internet, and to resolve short-range and mid-range protocol and architectural issues. It is a major source of proposals for protocol standards that are submitted to the IAB for final approval. The IETF meets three times a year and extensive minutes are included in the IETF Proceedings.

Internet Protocol (IP) The network layer for the TCP/IP Protocol Suite. It is a connectionless, best-effort, packet-switching protocol.

Internet Relay Chat (IRC) A world-wide "party line" protocol that allows you to converse with others in real time. IRC is structured as a network of servers, each of which accepts connections from client programs, one per user.

Internet Research Task Force (IRTF) The IRTF is chartered by the IAB to consider long-term Internet issues from a theoretical point of view. It has Research Groups (similar to IETF Working Groups) that are each tasked to discuss different research topics. Multi-cast audio/video conferencing and privacy-enhanced mail are samples of IRTF output.

Internet Society (ISOC)   The Internet Society is a nonprofit, professional membership organization that facilitates and supports the technical evolution of the Internet, stimulates interest in, and educates the scientific and academic communities, industry, and the public about the technology, uses, and applications of the Internet, and promotes the development of new applications for the system. The Society provides a forum for discussion and collaboration in the operation and use of the global Internet infrastructure. The Internet Society publishes a quarterly newsletter, the *Internet Society News,* and holds an annual conference, INET. The development of Internet technical standards takes place under the auspices of the Internet Society with substantial support from the Corporation for National Research Initiatives under a cooperative agreement with the US Federal Government.

Internetwork Packet eXchange (IPX)   Novell's protocol used by Netware. A router with IPX routing can interconnect LANs so that Novell Netware clients and servers can communicate.

interoperability   The ability of software and hardware on multiple machines from multiple vendors to communicate meaningfully.

IP address   The 32-bit address, usually represented in dotted decimal notation, that defines a node on the Internet.

Kermit   A popular file transfer protocol developed by Columbia University. Because Kermit runs in most operating environments, it provides an easy method of file transfer. Kermit is not the same as FTP.

layer   Communication networks for computers can be organized as a set of more or less independent protocols, each in a different layer (also called *level*). The lowest layer governs direct host-to-host communication between the hardware at different hosts; the highest consists of user applications. Each layer builds on the layer beneath it. For each layer, programs at different hosts use protocols appropriate to the layer to communicate with each other. TCP/IP has five layers of protocols (OSI has seven). The advantages of different layers of protocols is that the methods of passing information from one layer to another are specified clearly as part of the protocol suite, and changes within a protocol layer are prevented from affecting the other layers. This greatly simplifies the task of designing and maintaining communication programs.

listserv   An automated mailing list distribution system originally designed for the Bitnet/EARN network.

Local Area Network (LAN)   A data network intended to serve an area of only a few square kilometers or less. Because the network is known to cover only a small area, optimizations can be made in the network signal protocols that permit data rates up to 100Mb/s.

lurking   No active participation on the part of a subscriber to an mailing list or USENET newsgroup. A person who is lurking is just listening to the discussion. Lurking is encouraged for beginners who need to get up to speed on the history of the group.

mail bridge   A mail gateway that forwards electronic mail between two or more networks while ensuring that the messages it forwards meet certain administrative criteria. A mail bridge is simply a specialized form of mail gateway that enforces an administrative policy with regard to what mail it forwards.

mail gateway   A machine that connects two or more electronic mail systems (including dissimilar mail systems) and transfers messages between them. Sometimes the mapping and translation can be quite complex, and it generally requires a store-and-forward scheme, whereby the message is received from one system completely before it is transmitted to the next system, after suitable translations.

mail path   A series of machine names used to direct electronic mail from one user to another. This system of e-mail addressing has been used primarily in UUCP networks that are trying to eliminate its use altogether.

mail server   A software program that distributes files or information in response to requests sent via e-mail. Internet examples include Almanac and netlib. Mail servers also have been used in Bitnet to provide FTP-like services.

mailing list   A list of e-mail addresses, used by a mail exploder, to forward messages to groups of people. Generally, a mailing list is used to discuss a certain set of topics, and different mailing lists discuss different topics. A mailing list can be moderated. This means that messages sent to the list are actually sent to a moderator who determines whether or not to send the messages on to everyone else. Requests to subscribe to, or leave, a mailing list should always be sent to the list's "-request" address (e.g., ietf-request@cnri.reston.va.us for the IETF mailing list).

Metropolitan Area Network (MAN)   A data network intended to serve an area approximating that of a large city. Such networks are being implemented by innovative techniques, such as running fiber cables through subway tunnels. A popular example of a MAN is SMDS.

moderator   A person, or small group of people, who manage moderated mailing lists and newsgroups. Moderators are responsible for determining which e-mail submissions are passed on to list.

Multipurpose Internet Mail Extensions (MIME)   An extension to Internet e-mail that provides the ability to transfer nontextual data, such as graphics, audio, and fax.

National Institute of Standards and Technology (NIST)   The United States governmental body that provides assistance in developing standards. Formerly the National Bureau of Standards (NBS).

National Research and Education Network (NREN)   The realization of an interconnected gigabit computer network devoted to high-performance computing and communications (HIPPI) services.

National Science Foundation (NSF)   A U.S. government agency whose purpose is to promote the advancement of science. NSF funds science researchers, scientific projects, and infrastructure to improve the quality of scientific research. The NSFNET, funded by NSF, is an essential part of academic and research communications. It is a high-speed "network of networks" that is hierarchical in nature. At the highest level, it is a backbone network currently comprising 16 nodes connected to a 45Mb/s facility that spans the continental United States. Attached to that are mid-level networks, and attached to the mid-levels are campus and local networks. NSFNET also has connections out of the U.S. to Canada, Mexico, Europe, and the Pacific Rim. The NSFNET is part of the Internet.

netiquette   A pun on "etiquette" referring to proper behavior on a network such as the Internet.

network   A computer network is a data communications system that interconnects computer systems at various different sites. A network can be composed of any combination of LANs, MANs or WANs.

network address   The network portion of an IP address. For a class A network, the network address is the first byte of the IP address. For a class B network, the network address is the first two bytes of the IP address. For a class C network, the network address is the first three bytes of the IP address. In each case, the remainder is the host address. In the Internet, assigned network addresses are globally unique.

Network Operations Center (NOC)   A location from which the operation of a network is monitored. Additionally, this center usually serves as a clearinghouse for connectivity problems and efforts to resolve those problems.

Network Time Protocol (NTP)   A protocol that ensures accurate local time-keeping with reference to radio and atomic clocks located on the Internet. This protocol is capable of synchronizing distributed clocks within milliseconds over long time periods.

Online Computer Library Catalog (OCLC)   A nonprofit membership organization offering computer-based services to libraries, educational organizations, and their users. The OCLC library information network connects more than 10,000 libraries worldwide. Libraries use the OCLC System for cataloging, interlibrary loan, collection development, bibliographic verification, and reference searching.

Open Systems Interconnection (OSI)   A suite of protocols, designed by ISO committees, to be the international standard computer network architecture.

OSI Reference Model A seven-layer structure designed to describe computer network architectures and the way that data passes through them. This model was developed by the ISO in 1978 to clearly define the interfaces in multi-vendor networks, and to provide users of those networks with conceptual guidelines in the construction of networks.

Packet InterNet Groper (PING) A program used to test reachability of destinations by sending them an ICMP echo request and waiting for a reply. The term is used as a verb: "Ping host X to see if it is up!"

Packet Switch Node (PSN) A dedicated computer whose purpose is to accept, route, and forward packets in a packet switched network.

packet switching A communications paradigm in which packets (messages) are individually routed between hosts, with no previously established communication path.

Point-to-Point Protocol (PPP) The protocol provides a method for transmitting packets over serial point-to-point links. Used for dial-up Internet connections.

Post Office Protocol (POP) A protocol designed to allow single-user hosts to read mail from a server. There are three versions: POP, POP2, and POP3. Latter versions are not compatible with earlier versions.

Postal Telegraph and Telephone (PTT) Outside the USA, PTT refers to a telephone service provider, which is usually a monopoly, in a particular country.

postmaster The person responsible for taking care of electronic mail problems, answering queries about users, and other related work at a site.

protocol A formal description of message formats and the rules two computers must follow to exchange those messages. Protocols can describe low-level details of machine-to-machine interfaces (e.g., the order in which bits and bytes are sent across a wire) or high-level exchanges between allocation programs (e.g., the way in which two programs transfer a file across the Internet).

queue A backup of packets awaiting processing.

Remote Procedure Call (RPC) An easy and popular paradigm for implementing the client-server model of distributed computing. In general, a request is sent to a remote system to execute a designated procedure, using arguments supplied, and the result returned to the caller. There are many variations and subtleties in various implementations, resulting in a variety of different (incompatible) RPC protocols.

repeater A device that propagates electrical signals from one cable to another.

Reseaux Associes pour la Recherche Europeenne (RARE) European association of research networks.

Reseaux IP Europeenne (RIPE) A collaboration between European networks that use the TCP/IP protocol suite.

Reverse Address Resolution Protocol (RARP)   A protocol that provides the reverse function of ARP. RARP maps a hardware address to an Internet address. It is used primarily by diskless nodes when they first initialize to find their Internet address.

Round-Trip Time (RTT)   A measure of the current delay on a network.

router   A device that forwards traffic between networks. The forwarding decision is based on network layer information and routing tables, often constructed by routing protocols.

Serial Line IP (SLIP)   A protocol used to run IP over dial-up serial lines, such as telephone circuits or RS-232 cables, interconnecting two systems.

server   A provider of resources (e.g., file servers and name servers).

SIG   A special interest group.

signature   The three- or four-line message at the bottom of a piece of e-mail or a Usenet article that identifies the sender. Large signatures (over five lines) are generally frowned upon.

Simple Mail Transfer Protocol (SMTP)   A protocol used to transfer electronic mail between computers. It is a server-to-server protocol, so other protocols are used to access the messages.

Simple Network Management Protocol (SNMP)   The Internet standard protocol developed to manage nodes on an IP network. It is currently possible to manage wiring hubs, toasters, jukeboxes, etc.

subnet address   The subnet portion of an IP address. In a subnetted network, the host portion of an IP address is split into a subnet portion and a host portion using an address (subnet) mask.

Switched Multimegabit Data Service (SMDS)   An emerging high-speed datagram-based public data network service developed by Bellcore and expected to be widely used by telephone companies as the basis for their data networks.

Systems Network Architecture (SNA)   A proprietary networking architecture used by IBM and IBM-compatible mainframe computers.

T1   An AT&T term for a digital carrier facility used to transmit a DS-1 formatted digital signal at 1.544 megabits per second.

T3   A term for a digital carrier facility used to transmit a DS-3 formatted digital signal at 44.746 megabits per second.

TCP/IP Protocol Suite   Transmission Control Protocol over Internet Protocol. This is a common shorthand that refers to the suite of transport and application protocols that runs over IP.

TELENET   A public packet switched network using the CCITT X.25 protocols. It should not be confused with Telnet.

Telnet   The Internet standard protocol for remote terminal connection service.

TIA   An e-mail acronym for "Thanks In Advance."

TN3270 A variant of the Telnet program that allows you to attach to IBM mainframes and use the mainframe as if you had a 3270 or similar terminal.

token ring A token ring is a type of LAN with nodes wired into a ring. Each node constantly passes a control message (token) on to the next; whichever node has the token can send a message. Often, "Token Ring" is used to refer to the IEEE 802.5 token ring standard, which is the most common type of token ring.

topology A network topology shows the computers and the links between them. A network layer must stay abreast of the current network topology to be able to route packets to their final destination.

Transmission Control Protocol (TCP) An Internet-standard transport layer protocol. It is connection-oriented and stream-oriented, as opposed to UDP.

Trojan Horse A computer program that carries within itself a means to allow the creator of the program access to the system using it.

TTFN An e-mail acronym for "Ta-Ta For Now."

twisted pair A type of cable in which pairs of conductors are twisted together to produce certain electrical properties.

UNIX-to-UNIX CoPy (UUCP) This was initially a program run under the UNIX operating system that allowed one UNIX system to send files to another UNIX system via dial-up phone lines. Today, the term is more commonly used to describe the large international network that uses the UUCP protocol to pass news and electronic mail.

Usenet A collection of thousands of topically named newsgroups, the computers which run the protocols, and the people who read and submit Usenet news. Not all Internet hosts subscribe to Usenet, and not all Usenet hosts are on the Internet.

User Datagram Protocol (UDP) An Internet-standard transport layer protocol. It is a connectionless protocol that adds a level of reliability and multiplexing to IP.

virus A program that replicates itself on computer systems by incorporating itself into other programs that are shared among computer systems.

white pages The Internet supports several databases that contain basic information about users, such as e-mail addresses, telephone numbers, and postal addresses. These databases can be searched to get information about particular individuals. They are so named because they serve a function akin to the telephone book.

WHOIS An Internet program that allows users to query a database of people and other Internet entities, such as domains, networks, and hosts, kept at the DDN NIC. The information for people shows a person's company name, address, phone number, and email address.

Wide Area Information Servers (WAIS)   A distributed information service that offers simple, natural language input, indexed searching for fast retrieval, and a "relevance feedback" mechanism that allows the results of initial searches to influence future searches. Public domain implementations are available.

Wide Area Network (WAN)   A network, usually constructed with serial lines, which covers a large geographic area.

Winsock (Windows Socket)   A Windows API that lets Windows applications run on a TCP/IP network.

World Wide Web (WWW or W3)   A hypertext-based, distributed information system created by researchers at CERN in Switzerland. Users can create, edit, or browse hypertext documents. The clients and servers are freely available.

worm   A computer program that replicates itself and is self-propagating. Worms, as opposed to viruses, are meant to spawn in network environments. Network worms were first defined by Shoch & Hupp of Xerox in ACM Communications (March 1982). The Internet worm of November 1988 is perhaps the most famous; it successfully propagated itself on over 6,000 systems across the Internet.

WRT   An e-mail acronym for "With Respect To."

WYSIWYG   An acronym for "What You See is What You Get."

X.25   A data communications interface specification developed to describe how data passes into and out of public data communications networks. The CCITT and ISO approved protocol suite defines protocol layers 1 through 3.

X.400   The CCITT and ISO standard for electronic mail. It is widely used in Europe and Canada.

X.500   The CCITT and ISO standard for electronic directory services.

Yellow Pages (YP)   A service used by UNIX administrators to manage databases distributed across a network.

APPENDIX

A

Internet Service Providers

(Organized by Country)

INTERNET SERVICE PROVIDERS—*ANDORRA*

Calvacom
E-mail: scom1@calvacom.fr
Phone: 33-1-34 63 19 19
Fax: 33-1-34 63 19 48
URL: http://www.calvacom.fr/
Services: PPP

Civil engineering in cyberspace
E-mail: ceic@uplift.fr
Phone: +33 1 42 46 48 80
Fax: +33 1 40 23 01 30
URL: http://ceic.uplift.fr/
Services: Shell, Web servers, private cgi-bin

EUnet France
E-mail: contact@EUnet.fr
Phone: +33 1 53 81 60 60
Fax: +33 1 45 74 52 79
URL: http://www.EUnet.fr/
Services: UUCP, PPP V34, ISDN, LAN, LL, WWW—Individual and professional services

Europages
E-mail: freeinfo@europages.com (automated information)
E-mail: admin@europa.com
Phone: 1 53 77 54 00
Fax: 1 4289 34 73
URL: http://www.europages.com/

FRANCE-TEASER
E-mail: sales@teaser.fr (automated information)
E-mail: chrissy@FOUR.net
Phone: +331 4750-6248
Fax: +331 4750-6293
URL: http://www.teaser.fr/
Services: PPP, SLIP, CSLIP, WWW

Francenet
E-mail: info@email.francenet.fr (automated information)
E-mail: herve_sainct@email.francenet.fr
URL: http://www.fwi.com/
Services: Only provider in France that allows modem connection from outside Paris with no "killer" fees (others require you dial a number in Paris, where fare can be 10 times your local communication fare).

Gale Force Computing
E-mail: steve@monaco.mc
Phone: 93.50.20.92
Fax: 93.50.45.26
URL: http://www.monaco.mc/
Services: PPP, SLIP, sale of computer games, advertising, tourist information on the Riviera, the F1 Grand Prix statistics and results.

IBM Global Network
See entry under *United States.*

IMAGINET—France
E-mail: info@imaginet.fr (automated information)
E-mail: robin@imaginet.fr
Phone: (+33) (43) 38-10-24
Fax: (+33) (43) 38-42-62
URL: http://www.imaginet.fr/
Services: SLIP, E-mail, World Wide Web, Newsgroups, FTP, Telnet, Gopher, Archie

Internet-Way
E-mail: info@iway.fr
Phone: +33 1 41 43 21 10
Fax: +33 1 41 43 21 11
URL: http://www.iway.fr/
Services: PPP V34, 64 ISDN, LAN, 64 LL, 128 LL, 256 LL, WWW

MicroNet
E-mail: infos@micronet.fr
Phone: 40.59.46.68
Fax: 45.79.39.71
URL: http://www.micronet.fr
Services: SLIP/PPP

OLEANE
E-mail: info@oleane.net (automated information)
E-mail: oksales@oknet.com
Phone: +33 1 43 28 32 32
Fax: +33 1 43 28 46 21
URL: http://www.oleane.net/
Services: PPP V34, ISDN, LAN, LL (19.2,64,128,256 Kbs), X25

PacWan
E-mail: infos@pacwan.mm-soft.fr (automated information)
E-mail: mme@pacwan.mm-soft.fr
Phone: +33 42 93 42 93
Fax: +33 42 21 19 81
URL: http://www.mm-soft.fr/
Services: SLIP

WorldNet—France
E-mail: info@worldnet.net
Phone: 33 1 40 37 90 90
Fax: 33 1 40 37 90 89
URL: http://www.worldnet.net/
Services: SLIP, CSLIP, PPP

INTERNET SERVICE PROVIDERS—*ANTIGUA AND BARBUDA (THE CARIBBEAN)*

All America Cables and Radio, Inc.
Dominican Republic
E-mail: echava@aacr.net
Phone: (809) 221-3211
Fax: (809) 686-2385
URL: http://aacr.net/
Services: WWW, SLIP, PPP, Shell

Caribbean Internet Service, Corp.
E-mail: webmaster@caribe.net
Phone: (809) 728-3992
Fax: (809) 726-3093
URL: http://www.caribe.net/
Services: Shell, SLIP, PPP, Dedicated Services from 9.6Kbps up to T1

Caribbean Resources International Inc.
E-mail: admin@caribnet.net
Phone: (+1) (809) 431-0415
Fax: (+1) (809) 429-5903
URL: http://www.caribnet.net/
Services: Unlimited Time Dial Up PPP Access, Free User Web Pages
Free LINUX Shell Account and E-Mail Account with Access
Commercial Web Pages, Commercial Web Sites
MaGIC Database Site

Internet (Bermuda) Limited
E-mail: info@ibl.bm
Phone: (809) 296-1800
Fax: (809) 295-7269
URL: http://www.ibl.bm/
Services: PPP, 19.2K and 56K leased lines, POP3 Mail Service, WWW, Internet Consulting

TeleCom Plaza Corp.
E-mail: manuenc@moran.com
Phone: 386-1111
URL: http://www.moran.com/
Services: please call

INTERNET SERVICE PROVIDERS—*ARGENTINA*

Comint-ar
E-mail: jbarletta@usa.ar.net
Phone: 011-541-813-8706
URL: http://www.intr.net/comintar/
Services: SLIP, WWW, Web design/server, consulting.
General support of all internet matters. Connectivity and network design and feasibility studies to implement internet services.

SatLink S.A.
E-mail: soporte@satlink.com
Phone: +54-1-474-4512
Fax: +54-1-474-4512
URL: http://www.satlink.com/
Services: PPP, SLIP, CSLIP, UUCP, Leased lines, Domain Forward Web hosting

SiON Online Services
E-mail: info@sion.com.ar
Phone: +541 656-9195
Fax: +541 469-1335
URL: http://www.shout.net/
Services: UUCP, E-Mail (SLIP, PPP coming)

INTERNET SERVICE PROVIDERS—*AUSTRALIA*

Access One
E-mail: info@aone.net.au
Fax: (03) 9580-5581
URL: http://www.aone.net.au/
Services: PPP, V.FC/V.34/ISDN, Virtual Private Networks, Home Pages

AccServ
E-mail: sales@ppit.com.au
Phone: (03) 747-9823
Fax: (03) 747-8152
URL: http://home.acadia.net/
Services: Shell, UUCP

Advanced Internet Systems P/L
E-mail: enquiries@ais.com.au
Phone: +61 (3) 9800 3739
Fax: +61 (3) 9887 2756
URL: http://www.ais.com.au/
Services: Dialup, ISDN

APANA (ACT Region)
E-mail: act@apana.org.au
Phone: (06) 2824328
Fax: (06) 2824328
URL: http://www.act.apana.org.au/
Services: UUCP, Shell, SLIP, PPP, Permanent IP

Artsnet Australia
E-mail: suephil@peg.apc.org
Phone: (618) 376-1831
Fax: (618) 362-0876
URL: http://www.peg.apc.org/
~artsnet/
Services: SLIP, PPP, via Pegasus Networks (28.8K)

AUSNet Services Pty Ltd
E-mail: sales@world.net
Phone: +61 2 241-5888
Fax: +61 2 241-5898
URL: http://www.world.net/
Services: SLIP, CSLIP, PPP, ISDN (64 Kbit), Frame Relay

Ballarat NetConnect Pty Ltd
E-mail: info@netconnect.com.au
Phone: +61 53 322 140
Fax: +61 53 302 820
URL: http://www.netconnect.com.au/
Services: PPP, Shell, WWW, SLIP, ISDN or Dial Up ISDN

Brisbane Internet Technology
E-mail: admin@bit.net.au
Phone: +61 7 38416444
URL: http://www.bit.net.au/
Services: SLIP/PPP, permanent (dedicated) modem and ISDN, WWW advertising.

C Source Code Control Systems
E-mail: chas@csccs.com.au
Phone: +61 6 2921480
Fax: +61 6 2921985
URL: http://www.csccs.com.au/
Services: Async 28.8K dialup SLIP/PPP,
Permanemt async access, synchronous serial or ISDN permanent access, anon-FTP.
For registered users: WWW server, Unix
accounts, primary domain name server,
mail exchange records.

connect.com.au pty ltd
E-mail: connect@connect.com.au
Phone: +61 3 528 2239
Fax: +61 3 528 5887
URL: http://www.connect.com.au/
Services: UUCP, SLIP, PPP, Virtual Web
Service
PSTN (14.4k, 28.8k), ISDN (38.4k, 64k,
128k)

Cooee Communications
E-mail: sysop@cooee.com.au
Phone: +61 43 696224
URL: http://www.cooee.com.au/
Services: Shell, PPP, windows interface

Cynergy
E-mail: prooke@cynergy.com.au
Phone: +61 7 3357 1100
Fax: +61 7 3357 1199
URL: http://www.cynergy.com.au/
Services: WWW, PPP, IRC, Email, 28,800
modems across all racks

Data Research Associates, Inc.
E-mail: berit@dra.com
Phone: (+1) (314) 432-6634 Request
Berit Nelson
(+1) (314) 432-1100 Request Berit Nelson
(+1) (800) 325-0888 Request Berit Nelson
Fax: (+1) (314) 993-8927 ATTN: Berit
Nelson
URL: http://www.dra.com/

DIALix
E-mail: info@dialix.com
Phone: 1902 292 004 (Australia)
+61 2 9948 6995 (Intl)
Fax: +61 2 9948 8449
URL: http://www.DIALix.COM/
Services: Shell, UUCP, SLIP, PPP

Dynamite Internet
E-mail: harry@dynamite.com.au
Phone: +61 242-8644
Fax: +61 241-2408
URL: http://www.dynamite.com.au/
Services: PPP, SLIP, WWW, Shell

First Link Internet Services
E-mail: accounts@fl.com.au
Phone: 873-3577
Fax: 873-3297
URL: http://www.fl.com.au/
Services: Shell, SLIP/PPP, permanent IP
address available.

Geko
E-mail: sales@geko.com.au
Phone: 61 02 439-1999
Fax: 61 02 439-1919
URL: http://www.geko.com.au/
Services: Shell, SLIP, PPP, Dedicated lines,
ISDN

IBM Global Network
See entry under *United States*

iiNet Technologies Pty Ltd
E-mail: iinet@iinet.net.au
Phone: +61-9-322-7770
Fax: +61-9-322-6660
URL: http://www.iinet.net.au/
Services: Shell, UUCP, SLIP, PPP, leased
lines (64K through E1)

Informed Technology
E-mail: info@it.com.au
Phone: +61-9-245-2279
URL: http://www.it.com.au/
Services: SLIP, Shell, WWW

Intercom Pacific
E-mail: sales@ipacific.net.au
Phone: (02) 281-1111
Fax: (02) 281-2219
URL: http://www.ipacific.net.au/
Services: SLIP, PPP, ISDN, Network Con-
sulting, WWW Services, Security, LAN
Connectivity

InterConnect Australia Pty Ltd
E-mail: sales@interconnect.com.au
Phone: +61 3 9528 2239
Fax: +61 3 9528 5887
URL: http://www.interconnect.com.au/
Services: SLIP, PPP, WWW

Interline
E-mail: info@interline.com.au
Phone: (06) 231 3386
Fax: (06) 299 9062
URL: http://www.interline.com.au/
Services: Modem access to 28.8Kbps and
beyond

Internet Access Australia
E-mail: webmaster@iaccess.com.au
Phone: +61 3 576 4222
Fax: +61 3 563 7854
URL: http://www.iaccess.com.au/
Services: WWW, FTP, email, National net-
work inc. Permanent Connections, 28.8
to 128K lines, SLIP, PPP, UUCP ISDN
or High Speed modem links

Kralizec Dialup Internet System
E-mail: info@zeta.org.au
Phone: +61-2-837-1397
Fax: +61-2-837-3753
URL:
http://www.zeta.org.au/Guest/main.ht
ml/
Services: Intermittent, permanent,
volume-charged or flat rate

Microplex Pty Ltd
E-mail: info@mpx.com.au
Phone: (02) 438-1234
Fax: (02) 906 2326
URL: http://www.mpx.com.au/
Services: SLIP, PPP, leased lines (56K
through T1)

Mid North Coast Internet
E-mail: info@midcoast.com.au
Phone: 065 84 0345
Fax: 065 84 3955
URL: http://www.midcoast.com.au/
Services: SLIP/PPP, ISDN, WWW

Netro, Your Internet Connection
E-mail: info@netro.com.au
Phone: +61 2 876 8588
Fax: +61 2 876 6088
URL: http://www.netro.com.au/
Services: PPP, ISDN, WWW

Nettech AU Pty Ltd
E-mail: info@ah.net
Phone: 1300 30 1024
+61 6 254 8322
Fax: 1300 30 6235
URL: http://www.ah.net/

Northern Rivers Gateway
E-mail: drobinso@nrg.com.au
Phone: +61 66 216999
Fax: +61 66 224773
URL: http://www.nrg.com.au/
Services: Dialup 28.8K access, PPP,
Appletalk Remote Access (ARA), IPX &
UNIX shell access

OzEmail
E-mail: sales@ozemail.com.au
Phone: (02) 391 0480
Fax: (02) 437 5888
URL: http://www.ozemail.com.au/

Penrith Netcom Pty. Ltd.
E-mail: luke@pnc.com.au
Phone: +61 (47) 357000
Fax: +61 (47) 357000
URL: http://www.pnc.com.au/
Services: Shell, SLIP/PPP, WWW,
ISDN/dial-up ISDN

Spirit Networks Pty Ltd
E-mail: admin@spirit.net.au
Phone: +616 15 486 708
Fax: +616 285 1987
URL: http://www.spirit.net.au/
Services: SLIP, WWW, Virtual Internet
Server, ISDN dialup, dedicated links

Starway Corporation P/L
E-mail: hostmaster@starway.net.au
Phone: (+61) (03) 98032086
(+61) (06) 2929586 Starway Canberra
Fax: (+61) (03) 98876152
URL: http://www.starway.net.au/
Services: Casual and permanent SLIP/PPP
connections using PSTN or ISDN ser-
vices. News, ftp, Web, IRC, email

The Message eXchange TMXP
E-mail: elaine@tmx.com.au
Phone: (02) 550-4448
Fax: (02) 519 2551
URL: http://www.tmx.com.au/
Services: SLIP, PPP, ISDN

TMX, The Message eXchange
E-mail: elaine@tmx.com.au
Phone: 061 2 550 4448
Fax: 061 2 519 2551
URL: http://www.//tmx.com.au/
Services: SLIP, PPP and IP connections
are available.

World Reach Pty Ltd
E-mail: info@wr.com.au
Phone: (02) 436 3588
Fax: (02) 436 3998
URL: http://www.wr.com.au/
Services: SLIP/PPP dial-up with no time
charges, Web page publishing, UUCP
dial-up, Permanent connections, installa-
tions and consulting

Zip Australia Pty. Ltd.
E-mail: admin@zip.com.au
Phone: +61 2 482 7015
Fax: +61 2 482 7131
URL: http://www.zip.com.au/
Services: SLIP, PPP, Shell

INTERNET SERVICE PROVIDERS—*AUSTRIA*

alpin
Austrian Link to Progressive International Networking
E-mail: support@alpin.or.at
Phone: +43 (0)662 / 45 94 54
Fax: +43 (0)662 / 45 48 89 (Address to: alpin)
URL: http://alpin.or.at/home.html/
Services: PPP, Shell, hytelnet, WWW

EUnet EDV DienstleistungsGmbH
E-mail: info@Austria.EU.net
Phone: +43 1 3174969
Fax: +43 1 3106926
URL: http://www.Austria.EU.net/
Services: UUCP, SLIP, PPP, WWW, leased line IP, Frame Relay, Consulting

IBM Global Network
See entry under *United States.*

INS Informations- und Netzwerksysteme GmbH
E-mail: office@ins.at
Phone: +43-1-5230910-58
Fax: +43-1-5230910-24
URL: http://www.ins.at/
Services: SLIP, PPP, WWW, e-mail, FTP, leased line IP, dial-up IP, consulting, Web design, Internet marketing, ISDN, UUCP

LINK-ATU / Medienzentrum der Hochschuelerschaft an der TU Wien
E-mail: sysop@link-atu.comlink.apc.org
Phone: +43 1 586 1868
URL: http://www.iamerica.net/
Services: mail, news/ Z-NETZ, Zconnect, Janus

Net4You GmbH
E-mail: office@net4you.co.at
Phone: +43-4242-257367
Fax: +43-4242-257368
URL: http://www.net4you.co.at/
Services: PPP, WWW, EMail, FTP, Gopher, leased lines consulting & web-design, internet marketing

Online Store Europe
E-mail: info@onlinestore.com
Phone: (+41) (75) 373-6677
Fax: (+41) (75) 373-6660
URL: http://www.onlinestore.com/
Services: Shell accounts, Email, Usenet, Wais, UUCP, Ftp, Telnet, Gopher, IRC, Web co-location, SLIP/PPP, 2400bps dial-in 9600 bps dial-in, 14.4k bps dial-in, 28.8k bps dial-in ISDN dial-in, 56k leased line, ISDN leased line, T1 leased line, T2 leased line, T3 leased line Frame relay, ATM, SMDS, Wireless IP(NATEL D DATA 9600) Hardware, Software, Web site consulting, Systems Integration, Security Consulting

PING—Personal InterNet Gate Austria
E-mail: office@ping.at
Phone: +43-1-319 43 36
Fax: +43-1-310 69 27
URL: http://www.ping.at/
Services: SLIP, CSLIP, PPP, ISDN (64 Kbit), UUCP

simon media
E-mail: info@sime.com
Phone: +43-316-813 8240
Fax: +43-316-813 8246
URL: http://www.sime.com/
Services: PPP, WWW, E-mail, FTP, leased line IP, consulting, Web-design, Internet marketing

Vianet Austria Ltd
E-mail: sales@via.at
Phone: +43-1-5892920
Fax: +43-1-58929220
URL: http://www.via.at/
Services: UUCP, SLIP/CSLIP/PPP, ISDN, Frame Relay, WWW, VPN, Firewalling, Consulting

INTERNET SERVICE PROVIDERS—*BAHAMAS*

See entries under *Antigua and Barbuda (The Caribbean)*.

INTERNET SERVICE PROVIDERS—*BARBADOS*

See entries under *Antigua and Barbuda (The Caribbean)*.

INTERNET SERVICE PROVIDERS—*BELGIUM*

Arcadis
E-mail: support@arcadis.be
Phone: +32 2 534-1100
Fax: +32 2 534-1188
URL: http://www.arcadis.be/
Services: PPP, SLIP, UUCP, WWW

EUnet Belgium
E-mail: sales@Belgium.EU.net
Phone: +32-(0)16-23 60 99
Fax: +32-(0)16-23 20 79
URL: http://www.Belgium.EU.net/
Services: SLIP, CSLIP, PPP, ISDN (64 Kbit), Leased Lines

IBM Global Network
See entry under *United States*.

Infoboard Telematics
E-mail: ocaeymaex@infoboard.be
Phone: +32 2 475 22 99
Fax: +32 2 475 25 32
URL: http://www.ib.be/
Services: SLIP, PPP, ISDN, leased line, Web design and Web space rental

INnet
E-mail: support@inbe.net
Phone: +32 3 2814983
Fax: +32 3 2814985
URL: http://www.innet.net/
Services: PSTN, ISDN, LL

Interpac
E-mail: info@io.org
Phone: +32 2 646 60 00
Fax: +32 2 640 36 38
URL: http://www.interpac.be/
Services: PPP, PSTN, ISDN, Leased lines

Interpac Belgium SA/NV
E-mail: info@interpac.be
Phone: +32-(0)2-646 60 00
Fax: +32-(0)2-640 36 38
URL: http://www.interpac.be/
Services: PPP, ISDN (64 Kbit/s), Leased Lines, Web design & Web space rental, Firewalls consultancy

KnoopPunt Informatie
E-mail: support@knooppunt.be
Phone: +32 233 81 55
Fax: +32 233 73 43
URL: http://kikk.knooppunt.be/
Services: Shell, IP, MAIL, APC NEWS

Link Line
E-mail: admin@linkline.be
Phone: +32-2-644 25 13
Fax: +32-2-644 26 41
URL: http://www.linkline.be/
Services: Dial-up access PPP 28.8
modems, e-mail, news web (free page for
users), ftp (mirror of cica and info-mac),
IRC, free phone support day and evening
(in french, dutch or english), welcome
package containing user manuel, (choice
of 3 languages: dutch, french, english)
and licence paid softwares (Mac or Windows)

Netropolis Belgium
E-mail: info@netropolis.be
Phone: +32-2-649.36.93
Fax: +32-2-649.95.87
URL: http://www.netropolis.be/
Services: PPP, SLIP/CSLIP, teleconfer-
encing, shareware libraries, Web page
hosting, Entertainment Server

PING Belgium
E-mail: webmaster@ping.be
Phone: +32 56 615894
Fax: +32 56 615958
URL: http://www.ping.be/
Services: PPP, Webspace

INTERNET SERVICE PROVIDERS—*BERMUDA*

See entries under *Antigua and Barbuda (The Caribbean).*

INTERNET SERVICE PROVIDERS—*BRAZIL*

DGLNet Internet Provider
E-mail: mariowil@dglnet.com.br
Phone: 55-192-360058
Fax: 55-192-313446
URL: http://www.dglnet.com.br/
Services: SLIP/CSLIP/PPP, leased lines
of 64kb 128kb, e-mail only, Web site cre-
ation/design, firewalls, training for cor-
porations

MPS Informatica—Curitiba PR
E-mail: absy@mps.com.br
Phone: 55 41 322 4744
Fax: 55 41 224 6288
URL: http://www.mps.com.br/
Services: Internet access/information
Provider; Software development and con-
sulting

INTERNET SERVICE PROVIDERS—*BULGARIA*

EUnet Bulgaria
E-mail: postmaster@Bulgaria.EU.net
Phone: +359 52 259135
Fax: +359 52 234540
URL: http://www.eunet.bg/
Services: SLIP/PPP, dedicated Internet
access

INTERNET SERVICE PROVIDERS—*CANADA*

See entries under *United States.*

INTERNET SERVICE PROVIDERS—*CAYMAN ISLANDS*

See entries under *Antigua and Barbuda (The Caribbean).*

INTERNET SERVICE PROVIDERS—*CHILE*

Chilenet, the Chilean Internet Provider
E-mail: erodrigu@chilenet.cl
Phone: +56 2 678 4269
Fax: +56 2 689 5531
URL: http://nexus.chilenet.cl
Services: Shell, SLIP, UUCP

IUSATEL CHILE S:A:
E-mail: dweinste@iusanet.cl
Phone: 56-2-2469155
Fax: 56-2-2469164
URL: http://206.48.128.150/
Services: SLIP/PPP, leased lines 9.6KB–4MB, WWW services

Netup Ltda.
E-mail: info@netup.cl
Phone: +56-2-2510346
Fax: +56-2-2510347
URL: http://www.netup.cl/
Services: Dialup, SLIP, PPP, sync and async dedicated, ISDN, corporate WWW pages

RdC S.A.—CHILE, Latin America.
E-mail: info@rdc.cl
Phone: +56 2 696-2277
Fax: +56 2 635-1132
URL: http://mailnet.rdc.cl/
Services: SLIP, PPP

Reuna, Chilean Internet Provider
E-mail: wwwmaster@reuna.cl
Phone: 56-2-274 0403
Fax: 56-2-274 09 28
URL: http://www.reuna.cl/
Services: SLIP, Leased Lines 9.6k–4MB

INTERNET SERVICE PROVIDERS—*COSTA RICA*

TicoNet Costa Rica
E-mail: sysop@ticonet.co.cr
Phone: 011-506-290-3344
Fax: 011-506-290-3355
URL: http://www.ticonet.co.cr/
Services: E-mail, WWW server, Usenet News, Local forums

INTERNET SERVICE PROVIDERS—*CYPRUS*

Cyprus Telecommunications Authority (CYTA)
E-mail: help.desk@cyta.com.cy
Phone: +357 2 486633
URL: http://194.42.36.1/
Services: SLIP/PPP, leased lines, email services, newsgroups

EUnet Cyprus / Cylink Information Services Ltd.
E-mail: sales@Cyprus.EU.net
Phone: +357 2 369 114
Fax: +357 2 459 852
URL: http://www.cyprus.eu.net/
Services: SLIP/PPP, leased lines, UUCP, hosting, consulting, software and hardware

IBM GLOBAL NETWORK

See entry under United States.

INTERNET SERVICE PROVIDERS—*CZECHOSLOVAKIA*

Czech Educational and Scientific NETwork (CESNET)
E-mail: pkjensen@aci.cvut.cz
Phone: +42-2/2435-2974
+42-2/2435-5202
Fax: +42-2/2431-0271
URL: http://www.cesnet.cz/
Services: WWW, Direct Connection or Dial-In, Full Internet and User Support

EUnet Slovakia
E-mail: info@Slovakia.EU.net
Phone: +42 7 725 306
Fax: +42 7 728 462
URL: http://www.eunet.sk/
Services: UUCP, SLIP, PPP, IP/X.25, leased line, WWW

INTERNET SERVICE PROVIDERS—*DENMARK*

CyberCity
E-mail: response@cybercity.dk
Phone: +45 33339496
Fax: +45 33339406
URL: http://www.cybercity.dk/
Services: Full dial-in PPP Internet connection including 2MB Webpages and free Internet education in our cyberlab; Web housing operation; Web page design

Cybernet
E-mail: info@cybernet.dk
Phone: 33252282
Fax: 38882527
URL: http://www.cybernet.dk/
Services: Full Internet access including free web page

Danadata
E-mail: postmaster@danadata.dk
Phone: (+45) 70 10 80 80
Fax: (+45) 89 30 75 10
URL: http://www.danadata.dk/
Services: PPP connection for via 28.8KB modems and ISDN Network connections for companies Web space lease

Dansk Internet Adgang, DIA
E-mail: info@ia.dk
Phone: +45 7443 1220
Fax: +45 7442 2300
URL: http://www.ia.dk/
Services: SLIP, WWW, WWW design

IBM Global Network
See entry under United States.

UNI-C
E-mail: Michael.Reich@uni-c.dk
Phone: +45 35 82 83 55
Fax: +45 31 83 79 49
URL: http://www.uni-c.dk/
Services: Internet Service Classic (text), Internet Service 2000 (PPP), speeds up to 28.8 kb/s

INTERNET SERVICE PROVIDERS—*DOMINICA*

See entries under *Antigua and Barbuda (The Caribbean)*.

INTERNET SERVICE PROVIDERS—*DOMINICAN REPUBLIC*

See entries under *Antigua and Barbuda (The Caribbean)*.

INTERNET SERVICE PROVIDERS—*EAST TIMOR (INDONESIA)*

Bogor Internet
E-mail: mikesgd@indo.net.id
Phone: 62-251-316386
Fax: 62-251-326967
URL: http://www2.bonet.co.id/
Services: Home page provider

IndoInternet, PT. (INDONET)
E-mail: sales@indo.net.id
Phone: +62 21 470-2889
Fax: +62 21 470-2965
URL: http://www.indo.net.id/
Services: Shell, SLIP/PPP, WWW home pages

Indonesia Online Access—Indonesia
E-mail: sales@idola.net.id
Phone: 62-21-2302345
Fax: 62-21-2303883
URL: http://www.idola.net.id/
Services: Shell, SLIP/PPP

PT. Indo Internet—Indonesia
E-mail: sales@indo.net.id
Phone: +62 21 470-2889
Fax: +62 21 470-2965
URL: http://www.indo.net.id/
Services: Shell, menu, e-mail, SLIP, PPP

PT. Rahajasa Media Internet—Indonesia
E-mail: info@rad.net.id
Phone: +62-21 850-6788
Fax: +62-21 850-6744
URL: http://www.rad.net.id/
Services: Dedicated connection

RADNET
E-mail: riza@rad.net.id
Phone: 62-21-8506788
Fax: 62-21-8506744
URL: http://www.rad.net.id/
Services: SLIP, Web presence

INTERNET SERVICE PROVIDERS—*EQUADOR*

Ecuanex, (Quito, Ecuador)
E-mail: intercom@ecuanex.ec
Phone: 593-2-227-014
URL: http://www.echo-on.net/
Services: UUCP off-line service, FAX, Info-Servers

INTERNET SERVICE PROVIDERS—*EGYPT*

RitseCom
E-mail: tkamel@ritsec.com.eg
Phone: 20 2 3403538
Fax: 20 2 3412139
URL: http://ritsec_www.com.eg/
Services: SLIP, Terminal Emulation

INTERNET SERVICE PROVIDERS—*FINLAND*

Clinet Ltd
E-mail: clinet@clinet.fi
Phone: +358-0-437-5209
Fax: +358-0-455-5276
URL: http://www.clinet.fi/
Services: PPP, Leased

ComPart
E-mail: uffe@pcb.compart.fi
Phone: +358-0-506-3329
Fax: +358-0-506-3368
URL: http://www.compart.fi/
Services: SLIP, Frame Relay

EUnet Finland Oy
E-mail: helpdesk@eunet.fi
Phone: +358-0-4002060
Fax: +358-0-4784808
URL: http://www.eunet.fi/
Services: SLIP, PPP, UUCP, ISDN,
Leased Line, WWW-rental POP mailbox
service, high-speed (V.34) and high-
availability nationwide modem pool

Helsinki Media (Systems name: TIETOKONE Online)
E-mail: manty@online.tietokone.fi
URL: http://www.hol.gr/
Services: E-mail, news, shell interface

IBM Global Network
See entry under *United States.*

Net People
E-mail: helpdesk@netppl.fi
Phone: +358 81 5515 000
Fax: +358 81 5515 001
URL: http://www.netppl.fi/
Services: Shell, SLIP/PPP, service devel-
opment and consulting, Internet courses,
WWW advertising, development and
design, system administration and installa-
tion

INTERNET SERVICE PROVIDERS—*FRANCE*

Calvacom
E-mail: scom1@calvacom.fr
Phone: 33-1-34 63 19 19
Fax: 33-1-34 63 19 48
URL: http://www.calvacom.fr/
Services: PPP

Civil engineering in cyberspace
E-mail: ceic@uplift.fr
Phone: +33 1 42 46 48 80
Fax: +33 1 40 23 01 30
URL: http://ceic.uplift.fr/
Services: Shell, Web servers, private cgi-
bin

EUnet France
E-mail: contact@EUnet.fr
Phone: +33 1 53 81 60 60
Fax: +33 1 45 74 52 79
URL: http://www.EUnet.fr/
Services: UUCP, PPP V34, ISDN, LAN,
LL, WWW—Individual and professional
services

Europspages

E-mail: freeinfo@europages.com (automated information)
E-mail: admin@europa.com
Phone: 1 53 77 54 00
Fax: 1 4289 34 73
URL: http://www.europages.com/
Services: The European Business Directory provides data on 150,000 suppliers positioned for export from 25 European countries. This service is multilingual.

FRANCE-TEASER

E-mail: sales@teaser.fr (automated information)
E-mail: chrissy@FOUR.net
Phone: +331 4750-6248
Fax: +331 4750-6293
URL: http://www.teaser.fr/
Services: PPP, SLIP, CSLIP, WWW

Francenet

E-mail: herve_sainct@email.francenet.fr
URL: http://www.fwi.com/
Services: Only provider in France that allows modem connection from outside Paris with no "killer" fees (others require you dial a number in Paris, which fare can be 10 times your local communication fare).

Gale Force Computing

E-mail: steve@monaco.mc
Phone: 93.50.20.92
Fax: 93.50.45.26
URL: http://www.monaco.mc/
Services: PPP, SLIP, sale of computer games, advertising, tourist info on the Riviera, the F1 Grand Prix statistics and results

IBM Global Network

See entry under *United States.*

IMAGINET—France

E-mail: info@imaginet.fr (automated information)
E-mail: robin@imaginet.fr
Phone: (+33) (43) 38-10-24
Fax: (+33) (43) 38-42-62
URL: http://www.imaginet.fr/
Services: SLIP, E-mail, World Wide Web, Newsgroups, FTP, Telnet, Gopher, Archie

Internet-Way

E-mail: info@iway.fr
Phone: +33 1 41 43 21 10
Fax: +33 1 41 43 21 11
URL: http://www.iway.fr/
Services: PPP V34, 64 ISDN, LAN, 64 LL, 128 LL, 256 LL, WWW

MicroNet

E-mail: infos@micronet.fr
Phone: 40.59.46.68
Fax: 45.79.39.71
URL: http://www.micronet.fr
Services: SLIP/PPP

OLEANE

E-mail: info@oleane.net (automated information)
E-mail: oksales@oknet.com
Phone: +33 1 43 28 32 32
Fax: +33 1 43 28 46 21
URL: http://www.oleane.net/
Services: PPP V34, ISDN, LAN, LL (19.2,64,128,256 Kbs), X25 WWW—Professional and individual services

PacWan

E-mail: mme@pacwan.mm-soft.fr
Phone: +33 42 93 42 93
Fax: +33 42 21 19 81
URL: http://www.mm-soft.fr/
Services: SLIP

WorldNet—France
E-mail: info@worldnet.net
Phone: 33 1 40 37 90 90
Fax: 33 1 40 37 90 89
URL: http://www.worldnet.net/
Services: SLIP, CSLIP, PPP

INTERNET SERVICE PROVIDERS—*GERMANY*

ARANEA Internet Partner GmbH
E-mail: sales@ARANEA.net
Phone: +49 6101 5350-0
Fax: +49 6101 5350-99
URL: http://www.ARANEA.net/
Services: Netsurf, PPP, Dialup IP, Commercial ISDN, leased line, WWW

AXIS Information Systems GmbH
E-mail: info@axis.de (automated information)
E-mail: support@axis.de
Phone: (+49) (9131) 6913-50
Fax: (+49) (9131) 6913-49
URL: http://www.axis.de/
Services: IP over Modem (SLIP, PPP), IP over ISDN WWW-Servers, WWW<-->Database Solutions WWW-Pages

bbTT Electronic Networks GmbH
E-mail: stefan@b-2.de.contrib.net
Phone: +49 30 817 50 99
Fax: +49 30 817 69 76
Services: IP with SLIP or PPP by Modem (14.4–28.8kbps) or ISDN (64 kbps)

Birch Internet
E-mail: birch@birch.de
Phone: +49 30 42124500
Fax: +49 30 42124505
URL: http://www.birch.de/
Services: LAN ISDN Exclusive: Internet access for a LAN via leased line. LAN ISDN: Internet access for a LAN via bi-directional ISDN line. Einzelplatz Analog: Internet access for a stand-alone computer via modem.

Connect! Internetservices
E-mail: Tom@Connectde.net
Phone: +49 8179 5278
Fax: +49 8179 5375
URL: http://www.connectde.net/
Services: IP access via PPP, Dialup or leased line analog or ISDN, WWW-space and FTP-space

EUnet Germany
E-mail: sales@Germany.EU.net
Phone: +49 231 972 00
Fax: +49 231 972 1111
URL: http://www.Germany.EU.net/
Services: PersonalEUnet-single PPP, DialEUnet-LAN via TCP/IP InterE-Unet—leased line, EUnet Online Services

European Computer-Industry Research Centre (ECRC)
E-mail: internet@ecrc.de (automated information)
E-mail: admin@europa.com
Phone: +49 89 92699 100
Fax: +49 89 92699 170
URL: http://www.ecrc.de/
Services: ISDN 64 or 128 kbs; fixed lines 9.6kbs and larger

GTN
E-mail: ab@contrib.net
Phone: +40-203-306 1700
Fax: +49-203-306 1705
URL: http://www.gtn.com/
Services: SLIP

IBM Global Network
See entry under *United States.*

iMNet Gesellschaft fur internationale Kommunikation und Netzwerke mbH
E-mail: support@iM.Net
Phone: +49 931 6191191
Fax: +49 931 613330
URL: http://www.iM.Net/
Services: UUCP, SLIP, CSLIP, PPP, IP over ISDN, WWW

Kaiserslautern Internet Solution Service
E-mail: info@kiss.de
Phone: 0631-31662-0
URL: http://www.kiss.de/
Services: PPP, mail, news, WWW, leased lines (digital or analog)

Lemke & Fuerst GbR
E-mail: info@lf.net
Phone: +49 711 7189847
Fax: +49 711 7189848
URL: http://www.lf.net:81/
Services: SLIP, CSLIP, PPP, ISDN (64 Kbit)

Muc.De
E-mail: admin@muc.de
Phone: +49 89/3246830
Fax: +49 89/32468351
URL: http://www.muc.de/

NCS Network Communication Systems GmbH
E-mail: info@nord.de
Phone: +49 441 984989 0
Fax: +49 441 984989 1
URL: http://www.nord.de/
Services: PPP, ISDN (64 Kbit), Frame Relay

NDH—Netzwerkdienste Hoeger
E-mail: info@ndh.com
Phone: +49 2241 924510
Fax: +49 2241 924511
URL: http://www.ndh.com/
Services: UUCP, CSLIP/PPP via Modem or ISDN, WWW

NET Network Expert Team GmbH
E-mail: info@n-e-t.de
Phone: +49 711 9768921
Fax: +49 711 9768933
URL: http://www.n-e-t.de/
Services: UUCP, SLIP, CSLIP, PPP, ISDN (64 Kbit)

NetUSE GmbH
E-mail: gmbh@netuse.de
Phone: +49 431 642 300
Fax: +49 431 642 399
URL: http://www.netuse.de/
Services: SLIP, WWW, WWW creation, firewalling, emergency services, system administration

Netzwerk und Telematic GmbH, Geschaeftsbereich Xlink,
E-mail: sales@xlink.net
Phone: +49 721 96520
Fax: +49 721 9652 210
URL: http://www.xlink.net/
Services: IP over ISDN (64 Kbit), IP over leased line up to 2Mbps, UUCP, X.400 PSI

Noris Network
E-mail: support@noris.net
Phone: +49 911 9959621
Fax: +49 911 5980150
URL: http://info.noris.net/
Services: SLIP, PPP, ISDN, Leased lines,
WWW, DNS services, mail/news services,
Protocol conversion (Fidonet/ZConnect)

Online Store Europe
E-mail: info@onlinestore.com
Phone: (+41) (75) 373-6677
Fax: (+41) (75) 373-6660
URL: http://www.onlinestore.com/

Onlineservice Nuernberg
E-mail: admin@osn.de
Phone: +49-911-3781-100
Fax: +49-911-3781-999
URL: http://www.osn.de/
Services: Mail/News/IP with SLIP or
PPP by Modem or ISDN

PING e.V.
E-mail: vorstand@ping.de
Phone: +49-231-9791-0
URL: http://www.ping.de/
Services: SLIP, CSLIP, PPP (14.4-
28.8+ISDN), UUCP, Webspace

Point of Presence GmbH
E-mail: service@pop.de
Phone: +49 40 25 19 20 25
Fax: +49 40 25 81 94
URL: http://www.pcix.com/
Services: SLIP, Web sites (local + remote)

PROTEL Telekommunikations GmbH
E-mail: hotline@protel.de
Phone: +49-1805304127
Fax: +49-6122998190
URL: http://www.protel.de/
Services: SLIP, Analog and ISDN dial-up
accounts, much more (any Internet ser-
vice except UUCP)

SpaceNet GmbH
E-mail: info@space.net
Phone: +49 89 324683-0
Fax: +49 89 324683-51
URL: http://www.space.net/
Services: SLIP/PPP/UUCP/ISDN/ana-
logue

transnet Internet Services Leipzig
E-mail: info@leipzig.trans.net
Phone: +49 341 2518994
Fax: +49 341 4427342
URL: http://www.leipzig.trans.net
Services: Shell, SLIP/PPP, WWW, analog
and ISDN dialup, schooling, consulting

INTERNET SERVICE
PROVIDERS—*GHANA*

Chonia Informatica
E-mail: info@osagyefo.ghana.net
Phone: +233 21 669420
Fax: +233 21 669420
URL: http://rzunextbet1.unizh.ch/

INTERNET SERVICE PROVIDERS—*GREECE*

FORTHnet

E-mail: pr@forthnet.gr
Phone: +30 81 391200
Fax: +30 81 391201
URL: http://www.forthnet.gr/
Services: Dialup-IP, Dialup-UUCP, X25, Leased Line asynch 9600/Vfast, Leased Line synch 19200/64K/128K/2M

Hellas On Line S.A.

E-mail: webmaster@prometheus.hol.gr
Phone: 6203047
URL: http://www.hol.gr/

INTERNET SERVICE PROVIDERS—*GRENADA*

See entries under *Antigua and Barbuda (The Caribbean).*

INTERNET SERVICE PROVIDERS—*GUAM*

Guam Community College Computer Center

E-mail: michael@linette.guam.net
Phone: 011-671-735-5619
Fax: 011-671-734-8330
URL: http://linette.guam.net/

Kuentos Communications, Inc.

E-mail: pcarlson@kuentos.guam.net
Phone: +671 477-5750
+671-477-9109
Fax: +671-477-4218
URL: http://www.guam.net/
Services: SLIP/PPP, UUCP, News, FTP, Telnet, WWW, Gopher, IRC

INTERNET SERVICE PROVIDERS—*HONG KONG*

HKIGS

E-mail: helpdesk@hk.net
Phone: +852-2527-4888
Fax: +852-2527-4848
URL: http://www.hk.net/
Services: SLIP, CSLIP, PPP, 9.6K, 19.2K, 64K

Hong Kong Star Internet Limited

E-mail: info@hkstar.com
Phone: 2710-3209
Fax: 2770-8560
URL: http://www.hkstar.com/

Hong Kong Supernet

E-mail: info@HK.Super.NET
Phone: (852) 23587924
Fax: (852) 23587925
URL: http://www.hk.super.net/
Services: SLIP/PPP, 64K-256K leased line, WWW Hosting

Internet Access HK Limited

E-mail: info@ia.com.hk
Phone: (852) 2421 3121
Fax: (852) 24010055
URL: http://www.ia.com.hk/
Services: SLIP/PPP, 64K-T1 leased lines, Web rental, Web advertisement, training course

LinkAGE Online Limited

E-mail: help@hk.linkage.net
Phone: (852) 2331-8123
Fax: (852) 2795-1262
URL: http://www.hk.linkage.net/
Services: SLIP, PPP, leased lines (up to 512k)

Vision Network Limited
E-mail: info@vol.net
Phone: +852-2311-8855
Fax: +852-2311-8881
URL: http://www.vol.net/
Services: Shell, SLIP, PPP, WWW, Access to shareware/files

INTERNET SERVICE PROVIDERS—*HUNGARY*

EUnet Hungary
E-mail: sales@Germany.EU.net (automated information)
E-mail: info@Hungary.EU.net
Phone: +36 1 2698281
Fax: +36 1 2698288
URL: http://www.eunet.hu/
Services: UUCP, Shell, SLIP, PPP, full IP X.25, leased line

INTERNET SERVICE PROVIDERS—*ICELAND*

IntlS/ISnet
E-mail: isnet-info@isnet.is
Phone: +354 525 4747
Fax: +354 552 8801
URL: http://www.isnet.is/
Services: Standard ISP

SURIS/ISnet
E-mail: isnet-info@isnet.is
Phone: +354 569 4747
+354 525 4747
Fax: +354 552 8801
URL: http://www.isnet.is/
Services: Standard ISP

INTERNET SERVICE PROVIDERS—*INDIA*

Live Wire! BBS
E-mail:
support@lwbom.miqas2.fidonet.org
Phone: (91-22) 577-1111
Fax: (91-22) 578-7812
(91-22) 579-2416
URL: http://www.linkline.be/
Services: FidoNet, UUCP

STATUS INDIAGATE
E-mail: more@indigate.lod.com
Phone: +91-11-698-5111
Fax: +91-11-698-5111
URL: http://ssnet.com/

INTERNET SERVICE PROVIDERS—*INDONESIA*

Bogor Internet
E-mail: mikesgd@indo.net.id
Phone: 62-251-316386
Fax: 62-251-326967
URL: http://www2.bonet.co.id/

IndoInternet, PT. (INDONET)
E-mail: sales@indo.net.id
Phone: +62 21 470-2889
Fax: +62 21 470-2965
URL: http://www.indo.net.id/
Services: Shell, SLIP/PPP, WWW home pages

Indonesia Online Access—Indonesia
E-mail: sales@idola.net.id
Phone: 62-21-2302345
Fax: 62-21-2303883
URL: http://www.idola.net.id/
Services: Shell, SLIP/PPP

PT. Indo Internet—Indonesia

E-mail: sales@indo.net.id
Phone: +62 21 470-2889
Fax: +62 21 470-2965
URL: http://www.indo.net.id/
Services: Shell, menu, e-mail, SLIP, PPP

PT. Rahajasa Media Internet— Indonesia

E-mail: info@rad.net.id
Phone: +62-21 850-6788
Fax: +62-21 850-6744
URL: http://www.rad.net.id/
Services: Dedicated connection

RADNET

E-mail: riza@rad.net.id
Phone: 62-21-8506788
Fax: 62-21-8506744
URL: http://www.rad.net.id/
Services: SLIP, Web presence

INTERNET SERVICE PROVIDERS—*IRISH REPUBLIC*

Cork Internet Services, Ltd.

E-mail: kissanej@cis.ie
Phone: +353 (021) 277124
Fax: +353 (021) 278410
URL: http://www.cis.ie/
Services: Shell, UUCP, SLIP, PPP, ISDN, 64K

Eirenet

E-mail: ron@eirenet.net
Phone: +353 21 274141
Fax: +353 21 271635
URL:
http://www.ege.edu.tr/Ege/EgeNET/
Services: Dial-In SLIP, PPP, over PSTN, ISDN & leased lines available.

HomeNet

E-mail: info@HomeNet.ie
Phone: +353-1-6797355
Fax: +353-1-4546659
URL: http://www.HomeNet.ie/
Services: SLIP/PPP, WWW, E-mail, AFTP

IBM Global Network

See entry under *United States.*

Ireland On-Line

E-mail: sales@iol.ie
Phone: 353-1-8551739
Fax: 353-1-8551740
URL: http://www.iol.ie/
Services: Shell, SLIP, CSLIP, PPP, UUCP, ISDN, Leased Line Access, WWW, Consultancy

INTERNET SERVICE PROVIDERS—*ISRAEL*

ACTCOM—ACTive COMmunication Ltd.

E-mail: info@actcom.co.il
Phone: 972-4-676115
Fax: 972-4-676088
URL: http://www.actcom.co.il/
Services: Shell, SLIP/PPP, Email Usenet-only, UUCP

DataServe Internet services

E-mail: sales@datasrv.co.il
Phone: +972-3-6192727
Fax: +972-3-6199525
URL: http://www.datasrv.co.il/
Services: Shell, SLIP, PPP

elroNet
E-mail: info@elron.net
Phone: +9-72-4-545042
Fax: +9-72-4-551166
URL: http://www.elron.net/
Services: SLIP/PPP, WWW

FullFeed Communications of the Fox Valley
E-mail: sales@atw.fullfeed.com
Phone: (800) 840-8205
Fax: (414) 725-2610
URL: http://www.atw.fullfeed.com/

IBM Global Network
See entry under *United States*

IBM Israel
E-mail: sales@ibm.net.il (automated information)
E-mail: globalnetwork@info.ibm.com
Phone: +972-3-6978663
Fax: +972-3-6978115
URL: http://www.ibm.net.il/
Services: SLIP, WWW housing/authoring, leased lines, frame-relay, firewalling, security services

Kav Manche
E-mail: info@trendline.co.il (automated information)
E-mail: bobhitt@katslair.com
Phone: (972) 638-8222
Fax: (972) 638-8288
URL: http://204.180.8.1/
Services: CompUserve, SLIP, PPP

NetMedia
E-mail: avi@netmedia.co.il
Phone: +972 2 795-860
Fax: +972 2 793-524
URL: http://www.netmedia.co.il/
Services: PPP, SLIP, Email only, web design/hosting, list management, FTP

NetVision—elroNet assoc.
E-mail: info@netvision.net.il (automated information)
E-mail: gmbh@netuse.de
Phone: +972-4-550-330
Fax: +972-4-550-345
URL: http://www.netvision.net.il/
Services: PPP, WWW, FTP, Gopher, List-serv, Leased-Line, Frame-Relay

RDataServe Information Services
E-mail: support@datasrv.co.il
Phone: +972-3-619-2727
Fax: +972-3-619-9525
URL: http://www.datasrv.co.il/
Services: WWW, PPP, SLIP, Shell, Full advertising, Internet school, Providing direct Internet access to networks

Shani Technologies Ltd.
E-mail: support@shani.net
Phone: +972-3-6391288
+972-3-6478551
+972-9-461661
Fax: +972-3-6391287
URL: http://shani.net/
Services: PPP, SLIP, TERMINAL, E-Mail only, WWW authoring and servers, FTP site servers

INTERNET SERVICE PROVIDERS—*ITALY*

Abacom s.a.s
E-mail: glatzm@system.abacom.it
Phone: 39-(0)434-660911
Fax: 39-(0)434-660142
URL: http://www.abacom.it/
Services: SLIP, PPP, ISDN, dedicated lines

Agora' Telematica
E-mail: r.cicciomessere@agora.stm.it
Phone: +39-6-69917423
Fax: +39-6-69920123
URL: http://www.agora.stm.it/
Services: SLIP, BBS, E-Mail, Telnet, Gopher, FTP, WWW, IRC, Gateways to many other services

CSP/ALPcom Internet Services
E-mail: info@alpcom.it
Phone: +39 11 3187407
Fax: +39 11 4618 ext 212
URL: http://www.alpcom.it/
Services: HOST, SLIP, CSLIP, PPP, ISDN (64 Kbit), Frame Relay, X25

CyberNet S.r.l.—Internet Service Provider
E-mail: alfox@cybernet.it
Phone: [+39] (0)81 526 8989
Fax: [+39] (0)81 526 7678
URL: http://www.cybernet.it/

DSnet
E-mail: dsnet@dsnet.it
Phone: +39-51-521285
Fax: +39-51-522109
URL: http://www.dsnet.it/
Services: Shell, SLIP, PPP, ISDN, CDN (64kbit max), WWW

EVA Informatica
E-mail: pepis@eva.it
Phone: +39 70 765747
Fax: +39 70 782222
URL: http://www.eva.it/
Services: Shell, PPP, WWW & supports in Sardinia

EXE srl
E-mail: roberto@ba.dada.it
Phone: +39 +883 330000
Fax: +39 +883 37357
URL: http://www.ba.dada.it/
Services: Dial-up, import-export, engineering, project management, international business

FASTNET SRL
E-mail: info@fastnet.it
Phone: +39-71-2181081
Fax: +39-71-2181233
URL: http://www.fastnet.it

ferryboard
E-mail: direttore@ferryboard.com (automated information)
E-mail: sales@futuris.net
Phone: 337397772
Fax: 332510294
URL: http://www.ferryboard.com/

I.NET S.p.A.
E-mail: c-staff@inet.it
Phone: +39 2 26162261
Fax: +39 2 26110755
URL: http://www.inet.it/
Services: UUCP, SLIP, CSLIP, PPP, ISDN, Frame Relay, Leased lines

IBM Global Network
See entry under *United States.*

In.ternet—ITTC
E-mail: claudio@telnetwork.it
Phone: +39 337 584931
Fax: +39 522 303849
URL: http://www.imperium.net/
Services: Access Provider, Consulting,
Publishing Electronic Catalogue, Service,
Marketing Engineering

Interactive Electronic Design
E-mail: mc0003@mclink.it
Phone: +39-6-418921
URL: http://www.mclink.it/
Services: Dial-in SLIP/PPP, BBS with
Internet gateway (FTP, telnet, ping,
Lynx, etc), POP(2/3) for e-mail

internet force srl
E-mail: custserv@internetforce.com
Phone: 39210765
Fax: 39210119
URL: http://www.internetforce.com/
Services: Access to companies and private
people, fast access, italian design, food;
full internet, no limits

InterWare
E-mail: paolo@interware.it
Phone: +39-40-774488
Fax: +39-40-773311
URL: http:/www.interware.it/
Services: Consulting and training for
HTML and general networking topics.

INTESA SpA
E-mail: berga@it.ibm.com
Phone: +39 11 7090485
Fax: +39 11 7723341
URL: http://www.internext.com/
Services: SLIP

ItalNet
E-mail: elisa@italnet.it
Phone: +39 142 456566
Fax: +39 142 452193
URL: http://www.italnet.it/italweb/
Services: SLIP, PPP, access provider, con-
sulting, publishing, marketing.

ITnet
E-mail: info@IT.net
Phone: +39-10-6503-641
Fax: +39-10-6563-400
URL: http://www.IT.net/
Services: SLIP, PPP, ISDN, Frame Relay,
X.25

ITTC—In.Ternet Trade Center
E-mail: claudio@xmail.ittc.it
Phone: +39 522 383 023
+39 337 584931
Fax: +39 522 303 849
URL: http://www.ittc.it/
Services: Shell, SLIP/PPP, WWW, con-
sulting, electronic catalogue publishing,
marketing, engineering, training

**NETTuno service (by CINECA,
Interuniversity Computer Center)**
E-mail: staff@nettuno.it
Phone: +0039 51 6599423 / 6599411
Fax: +0039 51 6592581
URL: http://www.nettuno.it/
Services: SLIP, CSLIP, PPP, ISDN, Frame
Relay, Hardwired lines

OOD Organize Development Data
E-mail: mbrughi@odd.it
Phone: 8558209
Fax: 8558209
URL: http://www.odd.it/
Services: Internet Service Provider, Soft-
ware and CD-ROM Mastering

Ouverture Service srl
E-mail: Airwolf@outwest.com
Phone: +39-566-43026
Fax: +39-566-43268
URL: http://194.184.30.3/
Services: SLIP/PPP, ISDN

Panservice snc
E-mail: info@lt.flashnet.it
Phone: +39/773/240001
Fax: +39/773/240006
URL: http://www.lt.flashnet.it/
Services: PPP, UUCP, WWW, IP on
Leased Lines, BBS Dial Up with PPP
access, Installation and Configuration of
Un*x systems

PDnet
E-mail: remo@protec.it
Phone: +39 49 8686155
Fax: +39 49 8685661
URL: http://www.protec.it/
Services: SLIP,PPP, ISDN, CDN, CDA

Pro.Net. srl
E-mail: lbrizzi@pronet.it
Phone: +39 +6 6640385
Fax: +39 +6 6640384
URL: http://www.pronet.it/
Services: Connectivity on RTC, CDN,
CDA, ISDN. Hosting and Housing ser-
vices

RMnet Communications
E-mail: info@rmnet.it
Phone: +39 6 8530 2737
Fax: +39 6 8530 2737
URL: http://www.rmnet.it/
Services: PPP Dialup, shell accounts

**SPIN Internet Services, by Enter
s.a.s.(Trieste)**
E-mail: info@spin.it
Phone: +39 40 380700
+39 40 8992 286
Fax: +39 40 8992 257
URL: http://www.spin.it/
Services: SLIP/PPP, ISDN, leased lines,
WWW, consulting, Web
development/housing, training, seminars

Symbolic s.r.l.
E-mail: carla@symbolic.pr.it
Phone: +39 (0)521 221196
Fax: +39 (0)521 221099
URL: http://www.symbolic.pr.it/
Services: PPP, SLIP, WWW, Shell, E-mail

System srl—EZ-Net
E-mail: pbernad@eznet.it
Phone: +39-586-942732
Fax: +39-586-942205
URL: http://www2.eznet.it/
Services: SLIP, WWW

Tau S.r.l.
E-mail: Support@mail.tau.it
Phone: +39 95 7212146
Fax: +39 95 7212880
URL: http://www.tau.it/
Services: SLIP/PPP, dedicated lines,
WWW, FTP, E-Mail, BBS

Telnet S.r.l.
E-mail: staff@telnetwork.it
Phone: +39-382-529751
Fax: +39-382-528074
URL: http://www.telnetwork.it/
Services: SLIP

TEN s.a.s.
E-mail: staff@ten.iunet.it
Phone: +39 432 503982
Fax: +39 432 504083
URL: http://www.ten.iunet.it/
Services: PPP, Shell, Network services oriented to business and professional communities and www commercial site

TIZETAnet, Tizeta Informatica srl
E-mail: staff@tizeta.it
Phone: +39 51 346-346
Fax: +39 51 346-346
URL: http://www.tizeta.it/
Services: SLIP, PPP, ISDN (64 Kbit), X25

UniNet IP Services
E-mail: info@uni.net
Phone: +39 6 3938 7318
Fax: +39 6 3936 6949
URL: http://www.uni.net/
Services: RTC, ISDN (64 Kbit), CDN, CDA

Video On Line
E-mail: info@vol.it
Phone: +39-1670-14630
Fax: +39-1670-6013237
URL: http://www.vol.it/
Services: HOST, SLIP, CSLIP, PPP, CDN, ISDN, Frame Relay

INTERNET SERVICE PROVIDERS—*JAMAICA*

See entries under *Antigua and Barbuda (The Caribbean).*

INTERNET SERVICE PROVIDERS—*JAPAN*

Bekkoame Internet
E-mail: support@bekkoame.or.jp
Phone: 03 5610-7900
Fax: 03 5610-7901
URL: http://www.bekkoame.or.jp/
Services: Shell, PPP, UUCP, 64Kbps–1.5Mbps, ISDN (64K), Firewall

Global OnLine Japan
E-mail: sales@gol.com
Phone: +81-3-5330-9380
Fax: +81-3-5330-9381
URL: http://www.gol.com/
Services: Shell, PPP, ISDN, leased lines, WWW, Firewall Consulting

HA Telecom Internet Service
E-mail: info@hatelecom.or.jp
Phone: +81-58-253-7641
Fax: +81-58-253-7651
URL: http://www.hatelecom.or.jp/
Services: PPP, ISDN, leased lines, WWW, Consulting

IBM Global Network
See entry under *United States.*

Internet Initiative Japan, Inc. (IIJ)
E-mail: info@iij.ad.jp
Phone: +81-3-5276-6241
Fax: +81-3-5276-6239
URL: http://www.iij.ad.jp/
Services: UUCP, PPP, 64K, 128K, 384K, 512K, 768K, 1M, T1

ITJ Intelligent Telecom Inc. (ITJITNet)
E-mail: sales@itjit.ad.jp
Phone: +81-3-5565-3537
Fax: +81-3-5565-3940
URL: http://www.isp.net/isp_home.html/
Services: PPP, ISDN, Leased Lines (64K—2M)

Janis II
E-mail: pete.perkins@janis-tok.com
Phone: +81-3-3255-8880
Fax: +81-3-3255-8857
URL: http://www.entrepreneurs.net/ruth/
Services: PPP, 56K

RIMNET, (Rapid Systems CO.,Ltd.)
E-mail: pr@rim.or.jp
Phone: +81-3-5489-5655
Fax: +81-3-5489-5640
URL: http://www.rim.or.jp/
Services: PPP, dedicated line connection, UUCP

Twics
E-mail: twics@twics.com
Phone: +81 3 3351-5977
Fax: +81 3 3353-6096
URL: http://www.twics.com/
Services: Shell, PPP, UUCP, ISDN (64 Kbit)

Typhoon, Inc., Japan
E-mail: info@typhoon.co.jp (automated information)
E-mail: admin@typhoon.co.jp
Phone: (+81) (3) 3757-2118
Fax: (+81) (3) 3757-4640
URL: http://www.typhoon.co.jp/
Services: E-mail, PPP

INTERNET SERVICE PROVIDERS—*KUWAIT*

Gulfnet Kuwait
E-mail: info@kuwait.net
Phone: +965 242 6728
Fax: +965 243 5428
URL: http://www.kuwait.net/
Services: Shell, TIA, SLIP/PPP

INTERNET SERVICE PROVIDERS—*LIECHTENSTEIN*

CentralNet GmbH
E-mail: silvio@centralnet.ch
Phone: +41-41-20-30-50
URL: http://www.centralnet.ch/
Services: SLIP/PPP, leased lines (56K through T1)

EUnet AG, German speaking office of EUnet SA
E-mail: Admin@sparky1.arc.net
Phone: +41-1-291-45-80
+41-22-348-80-45
Fax: +41-1-291-46-42
URL: http://www.eunet.ch/
Services: SLIP, PPP, leased lines (64K through T2)

EUnet SA, French speaking office of EUnet AG
E-mail: deffer@eunet.ch
Phone: +41 22 348 80 45
Fax: +41 22 348 80 61
URL: http://www.eunet.ch/
Services: Leased lines SLIP/PPP over V34/ISDN, WWW space and consulting, NetNews, X.400 Mail

Fastnet Sarl
E-mail: info@fastnet.ch
Phone: (41 21) 324 06 76
Fax: (41 21) 324 06 74
URL: http://www.fastnet.ch/
Services: PPP, leased lines (64K through T1)

IBM Global Network
See entry under *United States.*

Internet Access AG
E-mail: webmaster@access.ch
Phone: +41 (1) 298-7777
Fax: +41 (1) 298-7776
URL: http://www.access.ch/
Services: PPP, ISDN, News Services, Web Hosting Services, Consulting

Internet ProLink
E-mail: info@iprolink.ch
Phone: +41-22-788-8555
Fax: +41-22-788-8560
URL: http://www.iprolink.ch/
Services: PPP Dial-up, leased lines, Web page hosting

Kern & Co.
E-mail: Mylene@keep.touch.ch
Phone: 0041 61 422 19 57
Fax: 0041 61 422 19 57
URL: http://www.touch.ch/
Services: SLIP, WWW, 24-hour support

Online Store Europe
E-mail: info@onlinestore.com
Phone: (+41) (75) 373-6677
Fax: (+41) (75) 373-6660
URL: http://www.onlinestore.com/

Ping Net
E-mail: admin@ping.li
Phone: 01 7358333
021 641 13 39
Fax: 01 7358334
URL: http://www.ping.ch/
Services: PPP, leased lines, WWW Page Rental, UUCP

Planet Communications SARL
E-mail: webmaster@planet.ch
Phone: +41-21-632-93-63
Fax: +41-21-632-93-64 20
URL: http://www.planet.ch/
Services: SLIP, PPP, leased lines

SWITCH—Swiss Academic & Research Network
E-mail: info@switch.ch
Phone: +41 1 268 1515
Fax: +41 1 268 1568
URL: http://www.switch.ch/
Services: SLIP, PPP, leased lines 64 kbit/s–2 Mbit/s

Worldcom
E-mail: admin@worldcom.ch
Phone: +41-21-802-51-51
Fax: +41-21-803-22-66
URL: http://www.worldcom.ch/
Services: SLIP, PPP, leased lines & ISDN (64K through 512K)

INTERNET SERVICE PROVIDERS—*LUXEMBOURG*

Europe Online S.A.
E-mail: inet-sales@eo.net
Phone: (+352) 40 101 226
Fax: (+352) 40 101 201
URL: http://www.eo.net/
Services: SLIP, PPP, leased lines (64k and up)

INTERNET SERVICE PROVIDERS—*MALAYSIA*

Jaring Net-Malaysia
E-mail: noc@jaring.my
Phone: (603) 254-9601
Fax: (603) 253-1898
URL: http://www.jaring.my/
Services: Most common internet services

INTERNET SERVICE PROVIDERS—*MEXICO*

Ashton Communications Corporation
E-mail: sales@acnet.net
Phone: (210) 668-6000
Fax: (210) 668-6001
URL: http://www.acnet.net/
Services: Latin America: dialup, dedicated 64-2048K (E0—E1), Web page hosting

Brisa Computacion Puebla
E-mail: brisa@noc.pue.udlap.mx
Phone: +52 (22) 40 70 93, 43 65 20
Fax: +52 (22) 37 55 63
URL: http://cyberware.web.com.mx/brisa2.html/
Services: Internet access in terminal and PPP modes. E-mail for travelers, when you visit Puebla, Mexico, you can use our office to read your e-mail.

CMACT
E-mail: info@mail.cmact.com
Phone: 378-0636
Fax: 378-0696
URL: http://www.cmact.com/
Services: SLIP, PPP, Dedicated lines (from 28k to T1), WWW Mail Gateways (MSMAIL, more gateways coming)

COMIMSA
E-mail: scolunga@mail.comimsa.com.mx
Phone: (52-84) 16-93-46
Fax: (52-84) 15-30-48
URL: http://www.comimsa.com.mx/
Services: Dialup SLIP, Dedicated Links, Web Hosting Services News Services

CyberNet S.A. de C.V.
E-mail: cvelarde@mailcybermex.net
Phone: 011-52-62-602272
Fax: 011-52-62-602273
URL: http://www.cybermex.net/
Services: SLIP/PPP, WWW, dedicated lines

Datanet
E-mail: help@data.net.mx
Phone: 525-1075400
Fax: 525-1075405
URL: http://www.data.net.mx/
Services: SLIP/PPP, leased-lines (from E0, 64kbps, to T1, 2.04 megabits), Web-site creation/space

Giga-Com S.A. de C.V.
E-mail: jpadres@mail.giga.com
Phone: +52-8-3366260
Fax: +52-8-3389203
URL: http://www.giga.com/
Services: WWW, SLIP, PPP, and dedicated conections

Info ABC
E-mail: abc@infoabc.com
Phone: 559 8315
559 8303
Fax: 559 4416
URL: http://www.infoabc.com/
Services: DIAL-UP, LAN/WAN, Web Server, Web Site and Pages Design, Support, Training and Seminars

INFOLINK S.A DE C.V.
E-mail: info@infolink.net.mx (automated information)
E-mail: sales@infolink.net.mx
Phone: +52-16-291583
Fax: +52-16-291586
URL: http://www.infolnk.net/
Services: DIAL-UP, PPP, Dedicated lines, Web hosting, Web Authoring, Design, Support, Training and Seminars, HTML and Java capabilities

Infosel
E-mail: flozano@cid.infl5
URL: http://www.it.com.au/
Services: Basic (no usenet), 28.8, PPP, SLIP

InterMex
E-mail: mena@mail.intermex.com.mx
Phone: (52-42) 18-48-95
Fax: (52-42) 18-48-95
URL: http://www.intermex.com.mx/
Services: SLIP (up to 28.8kb), leased lines (up to 64kb), publication of Web pages

INTERNET DE MEXICO, S.A. DE C.V.
E-mail: info@mail.internet.com.mx
Phone: +525 360-2931
Fax: +525 373-1493
URL: http://www.internet.com.mx/
Services: WWW, FTP, Telnet, E-Mail and SLIP accounts (28.8K)

Internet del Centro, SA de CV
E-mail: infoags@infosel.net.mx
Phone: (+52) (49) 12-66-88
Fax: (+52) (49) 12-31-88
Services: Dial up (PPP) and dedicated connections, page design and publication, seminars

Mundo Internet
E-mail: webmaster@kin.cieamer.conacyt.mx
Phone: 81-29-60 Ext. 227, 228
Fax: 81-29-23
URL: http://w3mint.cieamer.conacyt.mx/

Nexus Net
E-mail: equijano@tag.acnet.net
Phone: +52 (93) 520355
Fax: +52 (93) 520356
URL: http://nexusparc.acnet.net/
Services: SLIP/PPP, dedicated lines E0/DS0, Bandwith E1 (2.04 Mb), connected at ACbone (6 Mbps growing at 45 Mbps), WWW, WWW development

Soluciones Avanzadas de Redes
E-mail: info@sar.net
Phone: +52 +5 420-5900
Fax: +52 +5 420-5909
URL: http://www.sar.net/
Services: PPP, SLIP, Shell, Internet consulting, Web development and parking, and applications and equipment integration

SPIN: Sistema Profesional de Informacion—Tecnologia UNO CERO, S.A. de C
E-mail: jmatuk@spin.com.mx
Phone: +52(5)628-6200
+52(5)628-6210
Fax: +52(5)628-6220
URL: http://spin.com.mx/
Services: Shell, SLIP, BBS, private BBS areas, shareware, WWW, DNS, software development (especially medical)

Su Enlace al Mundo SA
E-mail: ernestoh@infosel.net.mx
Phone: 11-11-21 / 11-17-37
Fax: 11-11-21 / 11-17-37
URL: http://www.p3.net/
Services: PPP, 28.800 kbps dial up, commercial connections, home pages, seminars

Universidad Regiomontana A.C.
E-mail: longoria@mail.ur.mx
Phone: 52-8-342-5294
Fax: 52-8-344-3470
URL: http://www.ur.mx/
Services: SLIP, PPP, WWW, FTP

INTERNET SERVICE PROVIDERS—*MONACO*

See entries under *France*.

INTERNET SERVICE PROVIDERS—*MONTSERRAT*

See entries under *Antigua and Barbuda (The Caribbean)*.

INTERNET SERVICE PROVIDERS—*NEPAL*

Mercantile Communications Pvt. Ltd.
E-mail: mailmgr@mos.com.np
Phone: 977-1-220773
Fax: 977-1-225407
URL: http://www.mcn.org/
Services: UUCP e-mail, E-mail, Newsnet, WWW, Gopher, Archie, FTP, Lynx

INTERNET SERVICE PROVIDERS—*NETHERLANDS*

bART Internet Services
E-mail: info@bart.nl
Phone: +31-70-3455349
Fax: +31-70-3645062
URL: http://www.bart.nl/
Services: SLIP, PPP, leased lines (14k4 to 64k or more)

CyberConsult BV
E-mail: jim@cyber.nl
Phone: +31 306572219
Fax: +31 30 6567233
URL: http://www.cyber.nl/
Services: WWW, marketing, consulting, databases

EuroNet Internet bv
E-mail: info@euro.net
Phone: +31 20 625 6161
Fax: +31 20 625 7435
URL: http://www.euro.net/
Services: PPP or leased lines; WWW server space (1Mb free for users' home pages, Domain registration (Local Registry for for .nl)

Hobbynet
E-mail: info@hobby.nl
Phone: +31 36 5361683
URL: http://www.hobby.nl/
Services: SLIP, CSLIP

Holland Online Connection to the Internet
E-mail: sales@hol.nl
Phone: +31 70 4016943
Fax: +31 70 4013155
URL: http://www.hol.nl/
Services: SLIP/PPP, leased lines, WWW creation/space rental

IBM Global Network
See entry under *United States.*

Internet Access Foundation—The Netherlands
E-mail: info@iaf.nl
Phone: +31 15 566108 (in Dutch)
+31 5982 2720 (in English)
Fax: +31 15 566108
URL: http://www.iaf.nl/
Services: SLIP, ISDN (64 Kbit), UUCP, WWW services, consultancy and education

Internet Bausch
E-mail: info@atlanta.com
Phone: +31 (10) 413 7055
Fax: +31 (10) 413 7264
URL: http://www.ib.com/
Services: Shell, SLIPP, PPP, Web development

NederNet Online Services
E-mail: info@nedernet.nl
Phone: 31 2518 70207
URL: http://www.nedernet.nl/
Services: PPP, leased lines (28,800)

NetLand Internet Services
E-mail: raarts@netland.nl
Phone: +31 20 6943664
Fax: +31 20 5608411
URL: http://www.netland.nl/
Services: Shell, SLIP, PPP, WWW

Plex Elektronische Informatie
E-mail: hello@plex.nl
Phone: +31 77 37 35 33 5
URL: http://www.plex.nl/
Services: 1) TCP/IP via SLIP/CSLIP/PPP, (2) Software installation, (3) WWW design and implementation, (4) Turn-key Internet for businesses

Psyline
E-mail: postmaster@psyline.nl
Phone: +31 80 430194
URL: http://www.psyber.com/
Services: Dial-in, UUCP, subdomains

PublishNET Netherlands BV
E-mail: jan@publishnet.nl
Phone: 10305
Fax: 18354
URL: http://www.publishnet.nl/
Services: Dialup PPP Acccounts with private URL. Leased lines up to 64K/sec/Private domains possible WWW server, free Weekly E-zine. Webvertizing, TCP/IP networking consultancy/ education

Stichting DataWeb
E-mail: info@dataweb.nl
Phone: +31 70 3819218
Fax: +31 70 3835168
URL: http://www.dataweb.nl/
Services: SLIP, CSLIP, PPP, ISDN (64K), leased lines up to 2 Mbit

Stichting Knoware
E-mail: knoware@knoware.nl
URL: http://www.dataweb.nl/
Services: SLIP, PPP, Apple Macintosh support via ARA. FTP services, meetings

Xs4all
E-mail: helpdesk@xs4all.nl
Phone: +31 20 6200294
Fax: +31 20 6222753
URL: http://www.xs4all.nl/
Services: Unix, SLIP, PPP

zeelandnet.nl
E-mail: beheer@zeelandnet.nl
Phone: 01107-2062
Fax: 01107-2292
URL: http://www.zeelandnet.nl/
zldnet/
Services: SLIP PPP

INTERNET SERVICE PROVIDERS—*NETHERLANDS ANTILLES*

IBM Global Network
See entry under *United States.*

INTERNET SERVICE PROVIDERS—*NEW ZEALAND*

Efficient Software New Zealand Limited
E-mail: admin@es.co.nz
Phone: +64 3 4777474
Fax: +64 3 4777030
URL: http://www.es.co.nz/
Services: SLIP, UUCP

IBM Global Network
See entry under *United States.*

Internet Company of New Zealand
E-mail: ikuo@iconz.co.nz
Phone: 64-9-358 1186
Fax: 64-9-300 3122
URL: http://www.iconz.co.nz/
Services: SLIP, PPP, Shell, leased lines, WWW

Lynx Internet
E-mail: info@lynx.co.nz
Phone: (+64) 3 3790 568
Fax: (+64) 3 3654 852
URL: http://www.lyceum.com/
Services: PPP, Shell, UUCP, leased lines

PlaNet Manawatu Internet Services
E-mail: system@manawatu.gen.nz
URL: http://www.manawatu.gen.nz/

Wave Internet Services a division of CSE(NZ) Ltd
E-mail: info@wave.co.nz
Phone: +64-7-838-2010
Fax: +64-7-838-0977
URL: http://www.wave.co.nz/
Services: SLIP/PPP dial-in and leased-line connections, e-mail, telnet, FTP, News & WWW

INTERNET SERVICE PROVIDERS—*NICARAGUA*

UniComp
E-mail: kaz@upx.net (automated information)
E-mail: computo@uni.ni
Phone: (505)(2) 783142
URL: http://www.upx.net/
Services: SLIP, PPP, UUCP

INTERNET SERVICE PROVIDERS—*NORWAY*

ABS Systems
E-mail: walthero@absnet.no
Phone: 550043
Fax: 555057
URL: http://www.absnet.no/

BGNett
E-mail: drift@bgnett.no
Phone: +47 55104909
Fax: +47 55101712
URL: http://www.bgnett.no/
Services: Full PPP-Internet connection. All regular services included.

IBM Global Network
See entry under *United States.*

MultiNet AS
E-mail: multinet@multinet.no
Phone: +47 72 55 59 66
Fax: +47 72 55 73 38
URL: http://www.multinet.no/
Services: Shell, SLIP/PPP, UUCP, leased lines, Web services, consulting, modem sales

Oslonett
E-mail: oslonett@oslonett.no
Phone: +47 22 46 10 99
Fax: +47 22 46 45 28
URL: http://www.oslonett.no/
Services: Shell, SLIP, PPP, UUCP, ISDN

PowerTech Information Systems Inc.
E-mail: post@powertech.no
Phone: +47-2220-3330
Fax: +47-2220-0333
URL: http://www.powertech.no/
Services: Provides all types of connections, dial-up or leased lines, multi-port serial cards, Linux sales, etc.

Telepost
E-mail: info@teleport.ee
Phone: +47 800 800 01
URL: http://www.telepost.no/

Vestnett
E-mail: info@vestnett.no
Phone: 55543787
Fax: 55962175
URL: http://www.vestnett.no/
Services: SLIP, PPP, also leased line IP, WWW, UUCP, MS Mail gateway, courses and internet-related consulting service

INTERNET SERVICE PROVIDERS—*PAKISTAN*

Brain Computer Serivces
E-mail: basit@brain.com.pk
Phone: +92 42 541-4444
Fax: +92 42 758-1126
URL: http://singnet.com.sg/;tdbrains/
Services: UUCP, PPP

INTERNET SERVICE PROVIDERS—*PANAMA*

Servicio Nacional Internet S.A.
E-mail: ventas@pananet.com (automated information)
E-mail: soporte@pananet.com
Phone: +507-227-2222
Fax: +507-227-4612
URL: http://www.pananet.com/
Services: Internet access; ftp, Ghoper, news, WWW

INTERNET SERVICE PROVIDERS—*PARAGUAY*

Digital Electronics Laboratory (L.E.D.)
E-mail: info@ledip.py
Phone: +595 21 334 650
Fax: +595 21 310 587
URL: http://www.dsport.com/
Services: UUCP

INTERNET SERVICE PROVIDERS—*PERU*

Red Cientifica Peruana
E-mail: operador@rcp.net.pe
Phone: +51 1 445-5168
+51 1 445-9286
+51 1 445-5797
Fax: +51 1 444-7799
URL: http://www.rcp.net.pe/
Services: Shell, SLIP, PPP, WWW

INTERNET SERVICE PROVIDERS—*PHILIPPINES*

i-Way Services
E-mail: i-way@iphil.net
Phone: (632)6341963
Fax: (632)6341963
URL: http://i-way.iphil.net/
Services: Web advertisement, home page consultancy, computer sales

IPhil Communications Network, Inc.
E-mail: info@iphil.net (automated information)
E-mail: ingram@ipsnet.net
Phone: +63 2 893-9705
Fax: +63 2 893-9710
URL: http://www.iphil.net/
Services: UUCP, PPP and leased line connectivity, Home page hosting, System integration and consulting, Customized services

INTERNET SERVICE PROVIDERS—*POLAND*

ATM
E-mail: szeloch@ikp.atm.com.pl
Phone: +48 22 6106073
Fax: +48 22 6104144
URL: http://www.atm.com.pl/
Services: SLIP, PPP

Internet Technologies Polska
E-mail: plutecki@it.com.pl
Phone: +48 630-6049
Fax: +48 630-6011
URL: http://www.it.com.pl/
Services: SLIP, CSLIP, Internet training/consulting, dial-up up to 28.8 kbps, leased lines up to 128 kbps domain reservation, WWW services, modems and software

MAGNUM
E-mail: piotr@m1.com
Phone: +48 (42) 370662
Fax: +48 (42) 370735
URL: http://www.m1.com
Services: Shell, SLIP/PPP, mail accounts, UUCP, domain reservation, WWW, support

MALOKA
E-mail: sysop@gate.maloka.waw.pl
Phone: +48 22 6305004
+48 22 6220202
Fax: +48 22 6305004
URL: http://bbs.maloka.waw.pl/
Services: WWW, PPP, Shell, SLIP, local access to files, conferences

Polska OnLine
E-mail: webmaster@pol.pl
Phone: 6635086
Fax: 6635281
URL: http://www.pol.pl/

INTERNET SERVICE PROVIDERS—*PORTUGAL*

Esoterica
E-mail: info@esoterica.pt
Phone: +351 1 760 41 01
Fax: +351 1 716 56 51
URL: http://wwww.esoterica.pt/
Services: SLIP/PPP, UUCP, leased lines, ISDN (64 Kbits), Web space rental

EUnet Portugal (run by PUUG)
E-mail: info@puug.pt
Phone: +351 (1) 294 2844
Fax: +351 (1) 295 7786
URL: http://www.Portugal.EU.net/
Services: UUCP, SLIP, CSLIP, PPP, ISDN (64 Kbit), Frame Relay, Leased Lines

Telepac
E-mail: info@telebyte.com
Phone: +351 1 790 70 00
Fax: +351 1 790 70 01
URL: http://www.telepac.pt/
Services: SLIP, PPP

INTERNET SERVICE PROVIDERS—*PUERTO RICO*

See entries under *Antigua and Barbuda (The Caribbean).*

INTERNET SERVICE PROVIDERS—*ROMANIA*

EUnet Romania SRL
E-mail: info@Romania.EU.net
Phone: +40-1-312.6886
Fax: +40-1-312.6668
URL: http://www.Romania.EU.net/
Services: E-mail (UUCP/POP3), TCP/IP LL, SLIP, PPP

INTERNET SERVICE PROVIDERS—*SAINT KITTS NEVIS ANQUILLA*

See entries under *Antigua and Barbuda (The Caribbean).*

INTERNET SERVICE PROVIDERS—*SAINT LUCIA*

See entries under *Antigua and Barbuda (The Caribbean).*

INTERNET SERVICE PROVIDERS—*SAINT VINCENT AND THE GRENADINES*

See entries under *Antigua and Barbuda (The Caribbean).*

INTERNET SERVICE PROVIDERS—*SAN MARINO*

Abacom s.a.s
E-mail: glatzm@system.abacom.it
Phone: 39-(0)434-660911
Fax: 39-(0)434-660142
URL: http://www.abacom.it/
Services: SLIP, PPP, ISDN, dedicated lines

Agora' Telematica
E-mail: r.cicciomessere@agora.stm.it
Phone: +39-6-6991742
Fax: +39-6-69920123
URL: http://www.agora.stm.it/
Services: SLIP, BBS, E-Mail, Telnet,
Gopher, FTP, WWW, IRC, Gateways to
many other services

CSP/ALPcom Internet Services
E-mail: info@alpcom.it
Phone: +39 11 3187407
Fax: +39 11 4618 ext 212
URL: http://www.alpcom.it/
Services: HOST, SLIP, CSLIP, PPP, ISDN
(64 Kbit), Frame Relay, X25

CyberNet S.r.l.—Internet Service Provider
E-mail: alfox@cybernet.it
Phone: [+39] (0)81 526 8989
Fax: [+39] (0)81 526 7678
URL: http://www.cybernet.it/

DSnet
E-mail: dsnet@dsnet.it
Phone: +39-51-521285
Fax: +39-51-522109
URL: http://www.dsnet.it/
Services: Shell, SLIP, PPP, ISDN, CDN
(64kbit max), WWW

EVA Informatica
E-mail: pepis@eva.it
Phone: +39 70 765747
Fax: +39 70 782222
URL: http://www.eva.it/
Services: Shell, PPP, WWW & supports in
Sardinia

EXE srl
E-mail: roberto@ba.dada.it
Phone: +39 +883 330000
Fax: +39 +883 37357
URL: http://www.ba.dada.it/
Services: Dial-up, import-export, engi-
neering, project management, interna-
tional business

FASTNET SRL
E-mail: info@fastnet.it
Phone: +39-71-2181081
Fax: +39-71-2181233
URL: http://www.fastnet.it
Services: SLIP/PPP, leased lines (64K
through T1), ISDN, frame-relay leased
lines, Web hosting, virtual Web hosting,
consulting, training, domain name regis-
tration

ferryboard
E-mail: direttore@ferryboard.com (auto-
mated information)
E-mail: sales@futuris.net
Phone: 337397772
Fax: 332510294
URL: http://www.ferryboard.com/

I.NET S.p.A.
E-mail: c-staff@inet.it
Phone: +39 2 26162261
Fax: +39 2 26110755
URL: http://www.inet.it/
Services: UUCP, SLIP, CSLIP, PPP,
ISDN, Frame Relay, Leased lines

IBM Global Network
See entry under *United States.*

In.ternet—ITTC
E-mail: claudio@telnetwork.it
Phone: +39 337 584931
Fax: +39 522 303849
URL: http://www.imperium.net/
Services: Access Provider, Consulting,
Publishing Electronic Catalogue, Service,
Marketing, Engineering

Interactive Electronic Design
E-mail: mc0003@mclink.it
Phone: +39-6-418921
URL: http://www.mclink.it/
Services: Dial-in SLIP/PPP, BBS with
Internet gateway (FTP, telnet, ping,
Lynx, etc), POP(2/3) for e-mail

internet force srl
E-mail: custserv@internetforce.com
Phone: 39210765
Fax: 39210119
URL: http://www.internetforce.com/
Services: Access to companies and private
people, fast access, italian design, food;
full internet, no limits

InterWare
E-mail: paolo@interware.it
Phone: +39-40-774488
Fax: +39-40-773311
URL: http:/www.interware.it/
Services: Consulting and training for
HTML and general networking topics

INTESA SpA
E-mail: berga@it.ibm.com
Phone: 167018936 / +39 11 7090485
Fax: +39 11 7723341
URL: http://www.internext.com/
Services: SLIP

ItalNet
E-mail: elisa@italnet.it
Phone: +39 142 456566
Fax: +39 142 452193
URL: http://www.italnet.it/italweb/
Services: SLIP, PPP, access provider, con-
sulting, publishing, marketing

ITnet
E-mail: info@IT.net
Phone: +39-10-6503-641
Fax: +39-10-6563-400
URL: http://www.IT.net/
Services: SLIP, PPP, ISDN, Frame Relay,
X.25

ITTC—In.Ternet Trade Center
E-mail: claudio@xmail.ittc.it
Phone: +39 522 383 023
+39 337 584931
Fax: +39 522 303 849
URL: http://www.ittc.it/
Services: Shell, SLIP/PPP, WWW, con-
sulting, electronic catalogue publishing,
marketing, engineering, training

**NETTuno service (by CINECA,
Interuniversity Computer Center)**
E-mail: staff@nettuno.it
Phone: +0039 51 6599423 / 6599411
Fax: +0039 51 6592581
URL: http://www.nettuno.it/
Services: SLIP, CSLIP, PPP, ISDN, Frame
Relay, Hardwired lines

OOD Organize Development Data
E-mail: mbrughi@odd.it
Phone: 8558209
Fax: 8558209
URL: http://www.odd.it/
Services: Internet Sevice Provider, Soft-
ware and CD-ROM Mastering

Ouverture Service srl
E-mail: Airwolf@outwest.com
Phone: +39-566-43026
Fax: +39-566-43268
URL: http://194.184.30.3/
Services: SLIP/PPP, ISDN

Panservice snc
E-mail: info@lt.flashnet.it
Phone: +39/773/240001
Fax: +39/773/240006
URL: http://www.lt.flashnet.it/
Services: PPP, UUCP, WWW, IP on
Leased Lines, BBS Dial Up with PPP
access, Installation and Configuration of
Unix systems

PDnet
E-mail: remo@protec.it
Phone: +39 49 8686155
Fax: +39 49 8685661
URL: http://www.protec.it/
Services: SLIP, PPP, ISDN, CDN, CDA

Pro.Net. srl
E-mail: lbrizzi@pronet.it
Phone: +39 +6 6640385
Fax: +39 +6 6640384
URL: http://www.pronet.it/
Services: Connectivity on RTC, CDN,
CDA, ISDN. Hosting and Housing ser-
vices

RMnet Communications
E-mail: info@rmnet.it
Phone: +39 6 8530 2737
Fax: +39 6 8530 2737
URL: http://www.rmnet.it/
Services: PPP Dialup, shell accounts

**SPIN Internet Services, by Enter
s.a.s.(Trieste)**
E-mail: info@spin.it
Phone: +39 40 380700
+39 40 8992 286
Fax: +39 40 8992 257
URL: http://www.spin.it/
Services: SLIP/PPP, ISDN, leased lines,
WWW, consulting, Web
development/housing, training, seminars

Symbolic s.r.l.
E-mail: carla@symbolic.pr.it
Phone: +39 (0)521 221196
Fax: +39 (0)521 221099
URL: http://www.symbolic.pr.it/
Services: PPP, SLIP, WWW, Shell, Email

System srl—Ez-Net
E-mail: pbernad@eznet.it
Phone: +39-586-942732
Fax: +39-586-942205
URL: http://www2.eznet.it/
Services: SLIP, WWW

Tau S.r.l.
E-mail: Support@mail.tau.it
Phone: +39 95 7212146
Fax: +39 95 7212880
URL: http://www.tau.it/
Services: SLIP/PPP, dedicated lines,
WWW, FTP, E-Mail, BBS

Telnet S.r.l.
E-mail: staff@telnetwork.it
Phone: +39-382-529751
Fax: +39-382-528074
URL: http://www.telnetwork.it/
Services: SLIP

TEN s.a.s.
E-mail: staff@ten.iunet.it
Phone: +39 432 503982
Fax: +39 432 504083
URL: http://www.ten.iunet.it/
Services: PPP, Shell, Network services oriented to business and professional communities and www commercial site

TIZETAnet, Tizeta Informatica srl
E-mail: staff@tizeta.it
Phone: +39 51 346-346
Fax: +39 51 346-346
URL: http://www.tizeta.it/
Services: SLIP, PPP, ISDN (64 Kbit), X25

UniNet IP Services
E-mail: info@uni.net
Phone: +39 6 3938 7318
Fax: +39 6 3936 6949
URL: http://www.uni.net/
Services: RTC, ISDN (64 Kbit), CDN, CDA

Video On Line
E-mail: info@vol.it
Phone: +39-1670-14630
Fax: +39-1670-6013237
URL: http://www.vol.it/
Services: HOST, SLIP, CSLIP, PPP, CDN, ISDN, Frame Relay

INTERNET SERVICE PROVIDERS—*SINGAPORE*

Pacific Internet (Singapore) Pte Ltd
E-mail: info@pacific.net.sg
Phone: +65 1800-872-1455
Fax: +65 773-6812
URL: http://www.pacific.net.sg/
Services: PPP, leased lines, 19.2kbps to 256kbps, WWW, Web colocation

Singapore Telecom
E-mail: sales@singnet.com.sg
Phone: +65 730-8079
Fax: +65 732-1272
URL: http://www.singnet.com.sg/
Services: Personal IP, Full Access, Terminal, UUCP, WWW Regional Peering for ISPs.

INTERNET SERVICE PROVIDERS—*SOUTH AFRICA*

Aztec Information Management
E-mail: info@aztec.co.za
Phone: 27-21 4192690
Fax: 27-21 251254
URL: http://www.aztec.co.za/
Services: Shell, SLIP, PPP, UUCP, WWW, Leased Lines

Commercial Internet Services/Worldnet Africa
E-mail: info@cis.co.za
Phone: +27 12 841-2892
Fax: +27 12 841-3604
URL: http://www.cis.co.za/
Services: SLIP, CSLIP, PPP Leased Lines, Consulting

IBM Global Network
See entry under *United States.*

Internet Africa
E-mail: info@iafrica.com
Phone: +27 21 683 4370 (international)
Fax: +27 21 683 5778 (international)
URL: http://www.iafrica.com/iafrica/home.html/
Services: SLIP, PPP, leased lines sync (64K, 128K, 192K . . .) Frame Relay, X25, SMDS (2M). Async available.

**Internetworking Africa (Pty) Ltd,
(Trading as Internet Africa)**
E-mail: info@iafrica.com
Phone: +27 21 683-4370
Fax: +27 21 683-4695
URL: http://www.iafrica.com/iafrica/home.html/
Services: SLIP, PPP, UUCP, leased lines (async and 64k), Frame relay, X25, SMDS 2MB/s

PIPEX SA (Public Internet Provider Exchange)
E-mail: andrew@pipex-sa.net
Phone: +27 11 233-3200
Fax: +27 11 233-3223
URL: http://www.pipex-sa.net/
Services: Licenced Dial up Software, Dial Up Connections, Leased Lines, Analogue, Digital, ISDN, LAN, WAN and Internet Consultancy, Internet Security Systems

PiX
E-mail: doug@pix.za
Phone: +27-11-678-0097
Fax: +27-11-678-0091
URL: http://www.pix.za/
Services: PPP, SLIP Dial-up and leased lines

SANGONeT: South Africa's Nonprofit Information and Communications Network
E-mail: support@wn.apc.org
Phone: 27 -11 838 6943
Fax: 27 -11 492-1058
URL: http://www.sig.net/
Services: SLIP

SigNet Information Services
E-mail: cdg@cis.co.za
Phone: 27-31-732005
Fax: 27-31-732005
URL: http://africa.cis.co.za/
Services: SLIP, consulting, Web publishing, modem sales

The Internet Solution
E-mail: sales@is.co.za
Phone: +27 11 447-5566
Fax: +27 11 447-5567
URL: http://www.mainstreet.net/
Services: Shell, LAN

INTERNET SERVICE PROVIDERS—*SOUTH KOREA*

Dacom
E-mail: help@bora.dacom.co.kr
Phone: (+82) (2) 220-5207
(+82) (2) 220-5208
Fax: (+82) (2) 220-5329
URL: http://www.dacom.co.kr/
Services: Shell, SLIP/PPP, Dedicated lines

Korea PC Telecom
E-mail: help@hitel.kol.net
Phone: +82 2-513-2088
Fax: +82 2-513-2088
URL: http://www.libby.org/
Services: Shell, PPP, Leased Line

KORNET (Korea Telecom)
E-mail: helpme@kornet.nm.kr
Phone: +82 2 766 5900
Fax: +82 2 766 5903
URL: http://www.kornet.nm.kr/
Services: Shell, Dial-up SLIP/PPP, PSTN, Leased Line and Servers(ftp, archie, gopher, irc, NNTP, WWW)

KTNET(Korea Trade Network)
E-mail: help@ktb.net
Phone: 82-02-551-8512
Fax: 82-02-551-2268
URL: http://laputa.ktnet.co.kr/
Services: EDI, PC Communications, System Integration

INTERNET SERVICE PROVIDERS—*SPAIN*

bitMailer Online
E-mail: ventas@bitmailer.com
Phone: +34 1 402 15 51
Fax: +34 1 402 41 15
URL: http://www.bitmailer.com/
Services: PPP, Web creation and hosting, Internet security consulting, Leased lines, ISDN, Internet On Demand Access

Goya Servicios Telematicos—EUnet Spain
E-mail: info@world.net
Phone: +34 1 413 4856
Fax: +34 1 413 4901
URL: http://www.eunett.es/
Services: SLIP, CSLIP, PPP, ISDN (64 Kbit), Leased, Consultancy, Web services

IBM Global Network
See entry under *United States.*

INTERCOM S.T.A S.A.
E-mail: aserena@intercom.es
Phone: +34-3-5802846
Fax: +34-3-5805660
URL: http://www.intercom.es/
Services: Shell, PPP, Web server/rent, BBS

MANOL
E-mail: manol@ibm.net (automated information)
E-mail: sysop@gate.maloka.waw.pl
Phone: 34.72.52.60.61
Fax: 34.72.52.60.58
URL: http://bbs.maloka.waw.pl/

Ready Soft
E-mail: webmaster@readysoft.es
Phone: +34 77 33 09 86
Fax: +34 77 54 29 14
URL: http://www.readysoft.es/
Services: Dial-up access, Point-to-point, E-mail, FTP, Archie, Gopher, Wais, Veronica, Telnet, MUD, WWW, Hyper-G, WWW projects, Customer Training, HTML Page Design and Programming, CGI-Bin Programming

SARENET
E-mail: info@sarenet.es
Phone: +34 4 420 9470
Fax: +34 4 420 9465
URL: http://www.sarenet.es/
Services: SLIP/PPP, ISDN, frame-relay, leased lines, Web design and hosting

SERVICOM
E-mail: jhendler@servicom.es
Phone: 343 5809396
Fax: 343 5809098
URL: http://www.servicom.es/
Services: SLIP, PPP, leased lines (56K through T1)

INTERNET SERVICE PROVIDERS—*SRI LANKA*

Information Laboratories (PvT) Ltd,
E-mail: info@infolabs.is.lk
Phone: +94-1-61-1061
URL: http://www.is.lk/is/
Services: UUCP, E-mail, Usenet, FTP, BBS, Setting up Web pages

Lanka Internet Services, Ltd.
E-mail: info@lanka.net
Phone: +94-71-30469
Fax: +94-1-343056
URL: http://www.lanka.net/
Services: SLIP, PPP, leased lines (56K through T1)

INTERNET SERVICE PROVIDERS—*SVALBARD AND JAN MAYEN ISLANDS*

ABS Systems
E-mail: walthero@absnet.no
Phone: 550043
Fax: 555057
URL: http://www.absnet.no/

BGNett
E-mail: drift@bgnett.no
Phone: +47 55104909
Fax: +47 55101712
URL: http://www.bgnett.no/
Services: Full PPP-Internet connection. All regular services included.

IBM Global Network
See entry under *United States.*

MultiNet AS
E-mail: multinet@multinet.no
Phone: +47 72 55 59 66
Fax: +47 72 55 73 38
URL: http://www.multinet.no/
Services: Shell, SLIP/PPP, UUCP, leased lines, Web services, consulting, modem sales

Oslonett
E-mail: oslonett@oslonett.no
Phone: +47 22 46 10 99
Fax: +47 22 46 45 28
URL: http://www.oslonett.no/
Services: Shell, SLIP, PPP, UUCP, ISDN

PowerTech Information Systems Inc.
E-mail: post@powertech.no
Phone: +47-2220-3330
Fax: +47-2220-0333
URL: http://www.powertech.no/
Services: Provides all types of connections, dial-up or leased lines, multi-port serial cards, Linux sales, etc.

Telepost
E-mail: info@teleport.ee
Phone: +47 800 800 01
URL: http://www.telepost.no/

Vestnett
E-mail: info@vestnett.no
Phone: 55543787
Fax: 55962175
URL: http://www.vestnett.no/
Services: SLIP, PPP, also leased line IP, WWW, UUCP, MS Mail gateway, courses and internet-related consulting service.

INTERNET SERVICE PROVIDERS—*SWEDEN*

ABC-Klubben
E-mail: info@abc.se
Phone: +46 8 80 17 25
Fax: +46 8 80 15 22
URL: http://www.abc.se/
Services: A computer club with dial up access to Internet. Unix shell and SLIP or PPP. We use V.34+ USR Courier modem. We don't provide UUCP.

Aros InterNet AB
E-mail: the.crew@arosnet.se
Phone: +46 21 413360
Fax: +46 21 413360
URL: http://www.arosnet.se/
Services: WWW, PPP, POP, CSLIP, FTP, WWW-Motell

Bahnhof Internet Access—the Swedish station on infobahn
E-mail: bahnhof-info@bahnhof.se
Phone: +46 18 100899
Fax: +46 18 103737
URL: http://www.bahnhof.se/
Services: WWW, PPP, Shell, POP, SLIP, CSLIP

Canlt AB
E-mail: sales@canit.se
Phone: +46 (0)8-592 519 90
Fax: +46 (0)8-592 519 90
URL: http://www.canit.se/
Services: Shell, SLIP/CSLIP/PPP accounts, WWW hotel, UUCP

IBM Global Network
See entry under *United States.*

Lightning Line Service
E-mail: info@lls.se
Phone: +46-705-905526
Fax: +46-31-231753
URL: http://www.lls.se/
Services: Shell, PPP, WWW

NetG
E-mail: info@netg.se
Phone: +46-(0)31-280373
Fax: (201) 934-1445
URL: http://www.netg.se/
Services: PPP, UUCP, 19200–0.5 M, Website, Full kit

TerraTel
E-mail: info@netg.se
Phone: 46-(0)31-280373
URL: http://www.netg.se/
Services: PPP, UUCP, WWW, leased lines (19.2K to 0.5M)

INTERNET SERVICE PROVIDERS—*SWITZERLAND*

CentralNet GmbH
E-mail: silvio@centralnet.ch
Phone: +41-41-20-30-50
URL: http://www.centralnet.ch/
Services: SLIP/PPP, leased lines (56K through T1)

EUnet AG, German speaking office of EUnet SA
E-mail: info@eunet.ch (automated information)
E-mail: Admin@sparky1.arc.net
Phone: +41-1-291-45-80
+41-22-348-80-45
Fax: +41-1-291-46-42
URL: http://www.eunet.ch/
Services: SLIP, PPP, leased lines (64K through T2)

EUnet SA, French speaking office of EUnet AG
E-mail: deffer@eunet.ch
Phone: +41 22 348 80 45
Fax: +41 22 348 80 61
URL: http://www.eunet.ch/
Services: Leased lines SLIP/PPP over V34/ISDN, WWW space and consulting, NetNews, X.400 Mail

Fastnet Sarl
E-mail: info@fastnet.ch
Phone: (41 21) 324 06 76
Fax: (41 21) 324 06 74
URL: http://www.fastnet.ch/
Services: PPP, leased lines (64K through T1)

IBM Global Network
See entry under *United States.*

Internet Access AG
E-mail: webmaster@access.ch
Phone: +41 (1) 298-7777
Fax: +41 (1) 298-7776
URL: http://www.access.ch/
Services: PPP, ISDN, News Services, Web Hosting Services, Consulting

Internet ProLink
E-mail: info@iprolink.ch
Phone: +41-22-788-8555
Fax: +41-22-788-8560
URL: http://www.iprolink.ch/
Services: PPP Dial-up, leased lines, Web page hosting

Kern & Co.
E-mail: Mylene@keep.touch.ch
Phone: 0041 61 422 19 57
Fax: 0041 61 422 19 57
URL: http://www.touch.ch/
Services: SLIP, WWW, 24-hour support

Online Store Europe
E-mail: info@onlinestore.com
Phone: (+41) (75) 373-6677
Fax: (+41) (75) 373-6660
URL: http://www.onlinestore.com/

Ping Net
E-mail: admin@ping.li
Phone: 01 7358333
021 641 13 39
Fax: 01 7358334
URL: http://www.ping.ch/
Services: PPP, leased lines, WWW Page Rental, UUCP

Planet Communications SARL
E-mail: webmaster@planet.ch
Phone: +41-21-632-93-63
Fax: +41-21-632-93-64
URL: http://www.planet.ch/
Services: SLIP, PPP, leased lines

SWITCH—Swiss Academic & Research Network
E-mail: info@switch.ch
Phone: +41 1 268 1515
Fax: +41 1 268 1568
URL: http://www.switch.ch/
Services: SLIP, PPP, leased lines 64 kbit/s–2 Mbit/s

Worldcom
E-mail: admin@worldcom.ch
Phone: +41-21-802-51-51
Fax: +41-21-803-22-66
URL: http://www.worldcom.ch/
Services: SLIP, PPP, leased lines & ISDN (64K through 512K)

INTERNET SERVICE PROVIDERS—*TAIWAN*

Hinet
E-mail: nisc@hntp2.hinet.net
Phone: +886-2-3443143
URL: http://www.hinet.net/
Services: Shell, PPP, POPs, ISDN, leased lines

Institute for Information industry
E-mail: service@tpts1.seed.net.tw
Phone: 733-6454
733-8779
Fax: 737-0188
URL: http://www.seed.net.tw/
Services: SLIP, PPP, leased lines (9600 through T1)

Internet Solution Lab., Inc.
E-mail: jerome@isl.net.tw
Phone: +886-2-705-1835
Fax: +886-2-705-8601
URL: http://www.isl.net.tw/
Services: LAN-to-Internet, WWW

WOWNET Network Service Co., LTD.
E-mail: service@wow.net.tw
Phone: 998-3268
Fax: 998-5069
URL: http://www.wow.net.tw/
Services: PPP dial-up lines, Web page design, WWW space

INTERNET SERVICE PROVIDERS—*THAILAND*

INTERNET THAILAND SERVICE
E-mail: helpdesk@inet.co.th
Phone: (662) 642 7065
(662) 642 7066
Fax: (662) 642 7065
URL: http://www.inet.co.th/
Services: SLIP, leased lines (56K through T1)

KSC Commercial Internet Co. Ltd
E-mail: iru@ksc.net.th
Phone: 719-1948
Fax: 719-1945
URL: http://www.ksc.net.th/
Services: Leased lines 14.4-512K, dialup, Simulated SLIP/PPP, WWW pages

INTERNET SERVICE PROVIDERS—*TRINIDAD AND TOBAGO*

See entries under *Antigua and Barbuda (The Caribbean).*

INTERNET SERVICE PROVIDERS—*TURKEY*

EgeNET
E-mail: helpdesk@cakabey.ege.edu.tr
Phone: +90 232 3881080
URL:
http://www.ege.edu.tr/Ege/EgeNET/
Services: Dial-up, leased line services (both single and network connections), Shell/menu, SLIP/PPP accounts, 24-hour Internet service, help desk with software and hardware technical support

IBM Global Network
See entry under *United States.*

INTERNET SERVICE PROVIDERS— *UNITED KINGDOM*

AirTime Internet Resources Ltd UK
E-mail: sales@airtime.co.uk
Phone: +44 (0) 1254-676921
Fax: +44 (0) 1254-581574
URL: http://www.airtime.co.uk/
Services: SLIP, PPP, BBS

Aladdin
E-mail: info@aladdin.co.uk
Phone: +44 (0) 1489 782221
Fax: +44 (0) 1489 782382
URL: http://www.aladdin.co.uk/

avel PIP
E-mail: gary@avel.net
Phone: +44 (0)1872-262236
Fax: +44 (0)1872-262246
URL: http://www.avel.com/
Services: SLIP, PPP, FTP, WWW, Leased line, ISDN, Educational service, Consultancy

CityScape
E-mail: tony@ns.cityscape.co.uk
Phone: (01223) 566950
Fax: (01223) 566951
URL: http://www.citynet.net/
Services: Shell, SLIP, PPP

CIX (Compulink Information eXchange Ltd.)
E-mail: sales@compulink.co.uk
Phone: 0181 296 9666
Fax: 0181 296 9667
URL: http://www.compulink.co.uk/
Services: Conferencing (+text-based Internet access) system Dialup SLIP/PPP service

Colloquium
E-mail: McCarley@colloquium.co.uk
Phone: +44 141 849 0849-(international) 0141 849 0849-(national)
URL: http://www.colloquium.co.uk/
Services: Shell, PPP, IDSN lines available

Cyber Network Systems Limited
E-mail: chris@cyber.co.uk
Phone: +44-1925-494949
Fax: +44-1925-494950
URL: http://www.cyber.co.uk/
Services: ISDN, WWW, WWW design, POP3 mail, static IP address, Novell Network integration

Demon Internet Ltd
E-mail: sales@demon.net
Phone: +44-(0)181 371 1234
Fax: +44-(0)181 371 1150
URL: http://www.demon.co.uk/
Services: standard dial-up SLIP/PPP, 64K—2MB, WWW

Easynet
E-mail: admin@easynet.co.uk
Phone: (+44) 171 209 0990
Fax: (+44) 171 209 0891
URL: http://www.easynet.co.uk/
Services: SLIP, PPP, Web and FTP space, commercial Web services, leased lines via BTNet

EUnet GB Ltd
E-mail: sales@Britain.EU.net
Phone: +44 1227 266466
Fax: +44 1227 266477
URL: http://www.britain.eu.net/
Services: SLIP, PPP, 64K, 128K, 256K UUCP

FastNet International Ltd
E-mail: sales@fastnet.co.uk
Phone: +44 (1273) 677633
+44 (1273) 675314
Fax: +44 (1273) 621631
URL: http://www.fastnet.co.uk/
Services: SLIP, PPP, WWW, FTP, Usenet,
email, free home page, Internet training
courses

Foobar Internet
E-mail: sales@foobar.co.uk
Phone: +44 116 2330033
Fax: +44 116 2330035
URL: http://www.foobar.co.uk/
Services: SLIP, ISDN (64 Kbit), Web
pages, consultancy; 20:1 user to modem
ratio

Frontier Internet Services
E-mail: info@ftech.net
Phone: +44 (171) 242-3383
Fax: +44 (171) 242-3384
URL: http://www.ftech.net/
Services: SLIP, PPP, Web Space Rental

GreenNet
E-mail: support@gn.apc.org
Phone: +44 (0)171 713 1941
Fax: +44 (0)171 833 1169
URL: http://www.gn.apc.org/
Services: SLIP, PPP, email, conferences,
Web pages

Hiway, (Part of the Compass Computer Group)
E-mail: info@inform.hiway.co.uk
Phone: (44) (0)1635 550660
Fax: (44) (0)1635 521268
URL: http://www.hiway.co.uk/
Services: SLIP, PPP,(Dynamic or Static)
Dial-in (V34)

IBM Global Network
See entry under *United States.*

Mistral Internet
E-mail: info@mistral.co.uk
Phone: 01273 708866
URL: http://www.mistral.co.uk/
Services: Slip, PPP, 28.8K modems all lines

NETHEAD
E-mail: bob@nethead.co.uk
Phone: +44 171 207 1100
Fax: +44 171 207 5959
URL: http://www.nethead.co.uk/
Services: E-Mail (POP3 Recieve / SMTP
send) Full TCP/IP PPP Dialup connec-
tion, Web page hosting and design

Octacon Ltd
E-mail: onyx-support@octacon.co.uk
Phone: +44 (0) 1642 210 087
Fax: +44 (0) 1642 210 518
URL: http://www.octacon.co.uk/
Services: ISDN Dial-up, leased-lines,
WWW, Web installation and provision,
FTP services

On-line Entertainment Ltd
E-mail: mike@mail.on-line.co.uk
Phone: +44 0181 558 6114
Fax: +44 0181 558 3914
URL: http://on-line.co.uk/
Services: SLIP, WWW, Air Warrior,
Armoured Assault, Mud, Federation, TV-
Net, Forum, Britnet

Pavilion Internet pic
E-mail: sales@pavilion.co.uk
Phone: (+44) (1273) 607072
Fax: (+44) (1273) 607073
URL: http://www.pavilion.co.uk/

PC User Group/WinNET Communications

E-mail: info@win-uk.net
Phone: +44 181 863 1191
Fax: +44 181 863 6095
URL: http://www.ibmpcug.co.uk/
Services: UUCP, BBS, WWW Server and Client, SLIP, CSLIP, PPP

PIPEX (Public IP Exchange Limited) (part of Unipalm Group pic)

E-mail: sales@pipex.com
Phone: 0500 646566
+44(1223) 250120
Fax: +44(1223) 250121
URL: http://worldserver.pipex.com/
Services: Shell, SLIP, PPP, WWW, FTP, UUP, ISDN, Consulting

Planet Online

E-mail: sales@planet.net.au
Phone: 234 5566
Fax: 234 5533
URL: http://www.theplanet.co.uk/
Services: ISDN Internet Access for Businesses

Primex Information Services

E-mail: info@alpha.primex.co.uk
Phone: +44 (0)1908 313163
Fax: +44 (0)1908 643597
URL: http://www.primex.co.uk/
Services: SLIP, WWW

Pro-Net Internet Services

E-mail: sales@pro-net.co.uk
Phone: 0181 200 3565
Fax: 0181 205 8365
URL: http://www.pro-net.co.uk/
Services: Full IP connection using SLIP or PPP, WWW, FTP, email, London area only

Sound & Vision BBS (Worldspan Communications Ltd)

E-mail: rob@span.com
Phone: +44 0(181) 288 8555
Fax: +44 0(181) 288 8666
URL: http://span.com/
Services: BBS. Email. Usenet. FTP-by-Email. WWW-by-Email

Taynet

E-mail: sales@taynet.co.uk
Phone: +44 (0) 1382 561296
Fax: +44 (0) 1382 561444
URL: http://www.taynet.co.uk/
Services: SLIP, PPP (dynamic) Dial-up (V34 all lines) FTP/Web Space Rental

Technocom plc

E-mail: dguthrie@technocom.co.uk
Phone: +44 1753 673200
Fax: +44 1753 538415
URL: http://www.technocom.co.uk/
Services: SLIP, ISDN, leased lines, WWW, CD-ROMs online, LAN e-mail to Internet, online Web catalogue

The internet in nottingham (aka innotts)

E-mail: cindyc@innotts.co.uk
Phone: 0115 9562222
Fax: 0115 9561127
URL: http://www.innotts.co.uk/
Services: SLIP/PPP dial up for individuals, V34 modems, Web space rental, call for other services

Total Connectivity Providers Ltd
E-mail: sales@tcp.co.uk
Phone: +44 (0)1703 393392
Fax: +44 (0)1703 393392
URL: http://www.tcp.co.uk/
Services: SLIP, CSLIP, PPP, WWW/FTP
server space, domain registration, consultancy

U-NET Limited
E-mail: hi@u-net.com
Phone: +44 1925 633144
Fax: +44 1925 633847
URL: http://www.u-net.com/
Services: SLIP, PPP, ISDN, WWW Services

Wintermute Ltd.
E-mail: info@wintermute.co.uk
Phone: +44 (0) 1224 622477
Fax: +44 (0) 1224 637751
URL: http://www.wintermute.co.uk/
Services: SLIP, PPP, Menu/BBS Dial Up
access, Commercial Web Server, Consultancy, Provision of ISDN & Leased Line

World Business Net Ltd
E-mail: guyw@worldaccess.com
Phone: +44 1462 484345
Fax: +44 1462 484160
URL: http://www.flexnet.co.uk/business/
Services: Full access to Internet, Web
pages created, domains registered, leased
lines/ISDN lines installed, networks
installed

Zen Internet Ltd.
E-mail: sales@zenmail.co.uk
Phone: +44 1706 713714
Fax: +44 1706 175795
URL: http://www.zensys.co.uk/
Services: SLIP/PPP, Dial up/ISDN/Dedicated modem, WWW space, POP 3 mail,
Usenet news, domain registration, network consultancy, HTML training, Web
authoring

ZETNET Services
E-mail: info@zetnet.co.uk
Phone: +44 1595 696667
Fax: +44 1595 696548
URL: http://www.zetnet.co.uk/
Services: SLIP, WWW, FTP, Telnet, IRC

Zynet Ltd
E-mail: zynet@zynet.net
Phone: (+44)1392 426160
Fax: (+44)1392 421762
URL: http://www.zynet.co.uk/
Services: SLIP/PPP, FTP, WWW, Leased
line, ISDN, Educational service, consultancy

INTERNET SERVICE PROVIDERS—*UNITED STATES*

Arrownet
E-mail: info@arrownet.com (automated
information)
E-mail: the.crew@arosnet.se
Phone: (517) 371-7100
Fax: (517) 371-2188
URL: http://www.arrownet.com/
Services: PPP, WWW design, domain registration, consulting

BBN Planet Corporation
E-mail: net-info@bbnplanet.com
Phone: (800) 472-4565
Fax: (617) 873-3599
URL: http://www.bbnplanet.com/
Services: 19.2 Kbps, 56 Kbps, 128 Kbps,
256 Kbps, 384 Kbps, 768 Kbps, T1 (1.54
Mbps), 10 Mbps, T3 (45 Mbps)

Delphi
E-mail: sales@delphi.com
Phone: (800) 695-4005
(617) 491-3393
Fax: (617) 441-4903
URL: http://www.dash.com/
Services: Shell (online service)

EarthLink Network
E-mail: info@earthlink.net (automated
information)
E-mail: sales@earthlink.net
Phone: (818) 296-2400
(800) 395-8425
Fax: (818) 296-2470
URL: http://www.earthlink.net/
Services: Full Access to Internet with
TotalAccess Package includes the New
Netscape Navigator 2.0, Netscape mail,
and a FREE 2M Personal WEB Page

free.org
E-mail: info@free.org
Phone: (715) 743-1700
URL: http://www.ferryboard.com/
Services: Shell, SLIP, PPP

Global Enterprise Services, Inc.
E-mail: market@jvnc.net
Phone: (800) 358-4437x7325
(609) 897-7325
Fax: (609) 897-7310
URL: http://www.jvnc.net/
Services: Shell, SLIP, PPP, ISDN, 56K,
128K, 256K, 512k, T1+

Global Internet
E-mail: info@gi.net
Phone: (800) 682-5550
Fax: (402) 436-3036
URL: http://www.gi.net/
Services: 56k, 384k, T1, WWW, consult-
ing, 10MB ethernet ISP resell connections,
Internet & NT firewalls, Netra, VPNs

Greenlake Communications
E-mail: greenlak@cris.com
Phone: (810) 540-9380
Fax: (810) 540-0509
URL: http://www.cris.com/
~greenlak/index.html/
Services: SLIP/PPP, Web design, Web
advertising

HoloNet
E-mail: support@holonet.net
Phone: (510) 704-0160
Fax: (510) 704-8019
URL: http://www.holonet.net/
Services: Shell, SLIP, PPP, WWW dedi-
cated IP Service

Hypercon
E-mail: seans@hypercon.com
Phone: (800) 652-2590
Fax: (713) 995-9505
URL: http://www.hypercon.com/
Services: PPP, leased lines, LAN integra-
tion, software development, Web host-
ing/development, CGI-bin

IBM Global Network
E-mail: globalnetwork@info.ibm.com
Phone: (800) 775-5808
URL: http://www.ibm.com/
globalnetwork/
Services: Dial access (450 access points),
leased line access to T1, data security
arrangements, LAN-to-Internet connectivity, content management, Web hosting

Imagine Communications Corporation
E-mail: ceo@imagixx.net
Phone: (800) 5-MAGIXX
(800) 542-4499
(304) 292-6600 (NOC)
Fax: (304) 291-2577
URL: http://www.imagixx.net/
Services: SLIP/PPP, ISDN, dedicated
lines, 56K to T1, WWW, training, Web
development, LAN and WAN integration,
consulting

Institute for Global Communications
E-mail: support@igc.apc.org
Phone: (415) 442-0220
Fax: (415) 546-1794
URL: http://www.igc.apc.org/
Services: SLIP/PPP, telnet, FTP, gopher,
WWW, mailing lists, private conferences,
technical consulting

Internet Access Houston
E-mail: mike_st@iah.com
Phone: (713) 526-3425
Fax: (713) 522-5115
URL: http://www.iah.com/
Services: SLIP/PPP, ISDN, dedicated
lines, support services, Web page support

IPSnet
E-mail: ingram@ipsnet.net
Phone: (407) 426-8782
Fax: (407) 426-8984
URL: http://www.ipsnet.net/
Services: PPP, ISDN, UUCP, virtual host-
ing, secure server

Kallback
E-mail: zeke@kallback.com
Phone: (206) 286-5200
Fax: (206) 282-6666
URL: http://www.kallback.com/
worldnet/
Services: Shell, SLIP/PPP, Web site host-
ing, FTP servers, and more World-Wide
through unique call-back IP system

LogicalNET Corporation
E-mail: info@logical.net
Phone: (518) 452-9090
Fax: (518) 452-0157
URL: http://www.logical.net/
Services: SLIP, dedicated lines (to T3),
WWW

Northwest Internet Services, Inc
E-mail: marcotte@rio.com
Phone: (503) 342-8322
Fax: (503) 343-1699
URL: http://www.rio.com/
Services: Shell, SLIP, PPP, leased line
access, 56KB to T-1 web design services,
business support, network consulting,
association access plans

NovaLink Interactive Networks
E-mail: support@novalink.com
Phone: (800) 274-2814
URL: http://www.trey.com/
Services: Shell, SLIP

Portal Information Network
E-mail: sales@portal.com
Phone: (408) 973-9111
(800) 433-6444
Fax: (408) 725-1580
URL: http://www.portal.com/
Services: SLIP, PPP, UUCP, Shell

PSI (Performance Systems International)
E-mail: info@psi.com (automated information)
E-mail: hotline@protel.de
Phone: (800) 82PSI82
Fax: (800) FAXPSI1
URL: http://www.psi.com/
Services: Individual: Interramp, Pipeline USA, UUPSI, BBS UUCP, Sun Reseller Corporate: LAN Dial, LAN-ISDN, Inter-Frame, InterMAN, PSI Cable PSINET Services: Affinity Programs (for non-profit organizations), PSIWeb and Secure Connect

Questar Microsystems, Inc.
E-mail: gregp@questar.com
Phone: (206) 487-2627
(800) 925-2140
Fax: (206) 487-9803
URL: http://www.questar.com/
Services: PPP, ISDN, 56K thru T1, T3

Traders' Connection
E-mail: Lori@trader.com
Phone: (800) 753-4223
Fax: (317) 322-4310
URL: http://www.trader.com/
Services: host-dial, SLIP, PPP, T1, WWW

Unlearning Foundation
E-mail: lite@ix.netcom.com
Phone: (408) 423-8580
URL:
http://www.netcenter.com/air/air.html
Services: SLIP, 800 access when travelling

UUNET Technologies, Inc.
E-mail: info@uu.net (automated information)
E-mail: sales@uu.net
Phone: (800) 488-6384
(703) 206-5600
(+44) (1223) 250-100 UK
(+49) (231) 972-00 Germany
Fax: (703) 206-5601
URL: http://www.uu.net/
Services: Dedicated Access—56k through T3; Dial-Up Access—Analog and ISDN; Security Products and Services; Web and FTP Hosting Services; Consulting; Training; News; UUCP

Zocalo Engineering
E-mail: woody@zocalo.net
Phone: (510) 540-8000
Fax: (510) 548-1891
URL: http://www.zocalo.net/
Services: 56K, 128K, 384K, T1, Frame-Relay, AppleTalk Routing

ZONE One Network Exchange
E-mail: sales@zone.net
Phone: (212) 824-4000
(212) 824-5000
Fax: (212) 824-4009
URL: http://www.zone.net/
Services: UUCP, SLIP, PPP, 56K, T1, T3, ATM, FR, SMDS

INTERNET SERVICE PROVIDERS—*VENEZUELA*

Internet Comunicaciones, c.a.
E-mail: admin@ccs.internet.ve
Phone: +58-2-9599550
Fax: +58-2-9594550
URL: http://www.internet.ve/
Services: PPP dial-up

NetPoint Communications, Inc.
E-mail: rjc@netpoint.net
Phone: (305) 891-1955
Fax: (305) 891-2110
URL: http://www.netpoint.net/
Services: SLIP, PPP, ISDN, leased lines
(56K through T1)

INTERNET SERVICE PROVIDERS—*VIETNAM*

**NetNam Telematic Services, Vietnam
National Institute of Information Tech**
E-mail: admin@netnam.org.vn
Phone: +84-4 346-907
Fax: +84-4 345-217
URL: http://www.netmedia.co.il/
Services: UUCP, others in development

INTERNET SERVICE PROVIDERS—*VIRGIN ISLANDS*

See entries under *Antigua and Barbuda (The Caribbean)*.

INTERNET SERVICE PROVIDERS—*ZAMBIA*

**ZAMNET Communication Systems
Ltd, Lusaka, ZAMBIA**
E-mail: sales@zamnet.zm
Phone: +260 1 290358
+260 1 293317
Fax: +260 1 290358
URL: http://www.zamnet.zm/
Services: SLIP, PPP, E-Mail (rural users
only)

B

If you are on-line for any period of time, you will want to *download* files; that is, get files from on-line sources (i.e., drivers and utilities) and copy them into your computer for use. What you will find, however, is that software product files are typically packaged into compressed (sometimes called *archived*) files. In other words, the compressed file you download might actually contain many files related to the product you want. File compression serves several important purposes in the on-line world:

- Convenience   Large groups of files can be placed into a single compressed file. This means you can access an entire product by just downloading a single file instead of downloading all of the files you need individually.
- Speed   Compressed files are smaller (sometimes as small as 10% of the original, uncompressed form), so they can be downloaded faster. This saves telephone toll charges, as well as connect-time fees for your on-line service.

- Space Compression saves space on hard drives and network servers. This means that an on-line resource can fit more material into the available space.

The problem with compressed files is that they are not readily usable on a computer. Compressed files cannot be executed or read without first decompressing the file with a software utility. A "decompression" utility extracts each of the .EXE, .COM, .DAT, or other files contained in the compressed file, and places them in the current directory (or a sub-directory of your choice).

The exception to this is the "self-extracting" archive, which basically tags a copy of the decompression utility to the compressed file, then labels the file with an .EXE extension (appearing as an executable file). When you download a self-extracting file, simply execute it as you would any other program. This invokes the built-in decompressor, which decompresses the file automatically. The advantage of self-extracting files is their independence; you don't need a decompressor. The disadvantage is that the self-extraction code increases the compressed file size slightly, and takes away from some of the advantages of file compression.

The purpose of this appendix is to introduce you with the four most popular compression types, and show you where to obtain utilities needed to decompress those files. If you want detailed background or user information on these schemes, refer to the documentation that comes with each utility.

ARC

The ARC compression scheme (a registered trademark of System Enhancement Associates) produces compressed files with the .ARC extension. As one of the oldest PC compression schemes still in service, ARC is known for its good cross-platform compatibility with such systems as the Commodore Amiga. You also might find that ARC compression is well liked and broadly used in Europe, so keep an eye out for .ARC files when browsing European FTP or BBS sites.

There are a number of shareware and freeware ARC extractors available on-line. Try ARC-E.COM (and its associated document file ARC-E.DOC) located in Library 2 of the PCNEW Forum on CompuServe. You also can try the multi-purpose PAK251.EXE self-extracting utility, which is also in the PCNEW Forum.

ZIP

With a good balance between speed and compression, the ZIP scheme has become a de-facto standard for file compression in the US. Philip Katz's

PKWARE is one of the few shareware compression/decompression kits to make the successful transition to commercial software, and files compressed with PKZIP carry the .ZIP extension. Many of the BBS or other on-line resources that carry .ZIP files also carry a version of PKUNZIP available for download, though a growing number of third-party software makers have been developing multi-purpose compression/decompression utilities that are compatible with the .ZIP format (such as PAK251 or WINZIP).

You can choose from a number of ZIP-compatible decompression utilities in the PCNEW Forum on CompuServe. You also can order the complete PKWARE package to obtain the PKZIP, PKUNZIP, and ZIP2EXE (self-extractor) utilities.

ARJ

Files compressed with ARJ carry the .ARJ file extension. Robert Jung released the first version of ARJ in February of 1991. Since then, ARJ compression has become a popular alternative to ZIP compression. While ARJ compression runs a bit more slowly than other compression schemes, it achieves an excellent compression ratio, which is so vital for large file collections.

Like other compression schemes, there are a variety of decompression utilities available for download. The most popular utility is the self-extracting ARJ250.EXE shareware utility directly from Robert Jung. You can find this file (or a later version) in the PCNEW Forum of CompuServe. You also can get ARJ support information from the World Wide Web at http://www.dunkel. de/arj (German and English).

LHA

Originally known as the LH compression scheme, it was renamed LHA to avoid conflicts with the LH.EXE (loadhigh) feature incorporated into later versions of DOS. LHA compression creates files with the .LZH extension. LHA originated in Japan where it is a de-facto standard around the Pacific Rim. However, the LHA utility was released as free copyrighted software (useable without cost or royalty), so it has been gaining wide acceptance in the US commercial software industry where it can be included on distribution disks for free. LHA is one of the slower compression schemes, but it offers very good compression.

You can find the self-extracting LHA213.EXE utility in the PCNEW Forum on CompuServe, but multi-purpose compression/decompression utilities such as WINZIP also support LHA. For inquiries about LHA, you can contact K. Okubo on CompuServe at 74100,2565, or via the Internet at c00236@sinet.ad.jp.

GENERAL TOOLS

WINZip   (*handles ZIP, ARC, ARJ, LHA, gzip, and Unix compression*)
Nico Mak Computing, Inc.
PO Box 919
Bristol, CT 06011
Internet: http://www.winzip.com

PAK251.EXE   (*handles ARC, ZIP, and PAK compression*)
NoGate Consulting
P.O. Box 88115
Grand Rapids, MI 49518-0115
T: 616-455-6270
F: 616-455-8491
B: 616-455-5179

PKWARE, Inc.   (*handles ZIP compression*)
Philip Katz
9025 N. Deerwood Drive
Brown Deer, WI 53223
T: 414-354-8699
F: 414-354-8559
B: 414-354-8670
Internet: http://www.pkware.com
Email: sales@pkware.com

ARJ Software   (*handles ARJ compression*)
Robert Jung
P.O. Box 249
Norwood MA 02062 USA
F: 617-769-4893
B: 617-354-8873
Internet: robjung@world.std.com
CompuServe: 72077,445

C

Modem AT Command Set

M odems used to be "dumb" devices. It was almost impossible for them to do things like answer the ringing telephone line, dial a number, set speaker volume, and so on. Hayes Microcomputer Products developed a product called a "Smartmodem" which accepted high-level commands in the form of ASCII text strings. This was dubbed the *Hayes AT command set,* and has been the de-facto standard for modem commands ever since. As a consequence, virtually every modem that is *Hayes-compatible* is capable of using the AT command set. Ultimately, the AT commands go a long way to simplify the interface between a modem and communication software; rather than employ a cumbersome programming language, modem commands can be entered into any communications software, which are then passed to the modem when needed.

ON-LINE CONTROL COMMANDS

AT   A standard prefix that informs the modem a command is coming

A/   Repeat the previous command (this does *not* have to be prefixed by an AT)

+++ Switch modem from data mode to command mode

O0   Switch modem from command mode to data mode

DIALING AND ANSWERING COMMANDS

A   Answer the telephone manually

D   Dial the phone (usually used with a dial modifier code)

    P   *Pulse dial (simulate a rotary telephone)*

    R   *Originate the call and switch to answer mode (useful for originate-only modems)*

    T   *Tone dial (simulate a touch-tone telephone)*

    W   *Wait for a second dial tone (useful for dialing out of a PBX or other telco system)*

    ,   *Pause 2 seconds while dialing*

    !   *Flash the switch hook (simulate tapping the hook switch)*

    /   *Pause ⅛ second while dialing*

    @   *Wait for 5 seconds of silence before continuing (often requires 30 to 60 seconds)*

    ;   *Return modem to command mode after dialing*

H0   Hang up the telephone immediately

H1   Pick up the telephone immediately

CONTROLLING THE MODEM SPEAKER

L0   Set very low speaker volume

L1   Set low speaker volume

L2   Set medium speaker volume

L3   Set high speaker volume

M0   Shut off the speaker

M1   Disable speaker when connection is established

M2   Turn on speaker

CONTROLLING RESULT CODES

C0   Return an error response

C1   Use normal transmit carrier switching

Q0   Go ahead and send result codes
Q1   Do not send result codes
V0   Select numeric result codes
V1   Select text result codes

GENERAL MODEM COMMANDS

B0   Use the modem with ITU protocols
B1   Use the modem with Bell protocols
E0   Disable the command echo
E1   Enable the command echo
F0   Select half-duplex operation
F1   Select full-duplex operation
H0   Force the modem to hang-up
H1   Force the modem off hook
N0   Make connections at the speed specified in S37
N1   Make a connection at any available speed
X0   Do not wait for dial tone before dialing and do not report busy signals
X1   Do not wait for dial tone before dialing and do not report busy signals
       (extra result codes)
X2   Wait for a dial tone before dialing, but do not report busy signals
X3   Do not wait for a dial tone before dialing, but report busy signals
X4   Wait for a dial tone before dialing, and report busy signals
Y0   Disable the long space disconnect
Y1   Enable the long space disconnect
Z   Reset the modem to its power up defaults

AMPERSAND COMMANDS

&C0   Always assert the DCD signal
&C1   Only assert DCD while connected to a remote modem
&D0   Ignore the state of the DTR signal
&D1   Switch from data mode to command mode when DTR switches
        from on to off
&D2   Hand up and switch from data mode to command mode when
        DTR switches off
&D3   Reset the modem when DTR switches from on to off
&F   Restore the factory default settings
&J0   System uses RJ11 telephone jack
&J1   System uses RJ12 or RJ13 telephone jack
&K0   Disable flow control
&K3   Enable RTS/CTS flow control

&K4   Enable XON/XOFF flow control
&K5   Enable transparent XON/XOFF control
&L0   Using a dial-up telephone line
&L1   Selects a leased telephone line
&Q0   Use asynchronous operation in the direct mode
&Q1   Use synchronous mode 1
&Q2   Use synchronous mode 2
&Q3   Use synchronous mode 3
&Q5   Modem negotiates error correcting protocol
&Q6   Asynchronous operation in normal mode
&R0   Assert CTS only when RTS is asserted
&R1   Always assert CTS signal
&S0   Always assert DSR signal
&S1   Assert DSR only when in on-line mode
&W   Save current configuration settings in NVRAM
&Z   Reset the modem and recall settings from NVRAM

S-REGISTER ASSIGNMENTS

S0   Number of rings to wait before answering a call
S1   Check signal quality
S2   Escape code character (ASCII 43)
S3   Carriage return character (ASCII 13)
S4   Line feed character (ASCII 10)
S5   Backspace character (ASCII 8)
S6   Number of seconds to wait for dial tone before dialing
S7   Number of seconds to wait for carrier connection before hanging up
S8   Pause time represented by comma
S9   Length of time a carrier must be present before modem will recognize carrier
S10   Delay cutoff after carrier loss
S11   Duration of time between each touch tone signal when dialing
S12   Escape code guard time

FORMING A MODEM COMMAND

Virtually all AT command strings start with the prefix "AT" (Attention). For example, the command string **ATZE1Q0V1** contains five separate commands; attention (AT), reset the modem to its power-up defaults (Z), enable the command echo to send command characters back to the sender (E1), send command result codes back to the PC (Q0), and select text result codes, which causes words to be used as result codes (V1). While this might seem like a mouthful, a typical modem can accept command strings up to 40 characters long. The *result codes* are the messages that the modem generates when a command string is processed. Either numbers (default) or words (using the V1 command) can be returned. When a command is processed correctly, a result code "OK" is produced by the modem, or CONNECT when a successful connection is established.

THE ROLE OF COMMAND STRINGS

If you start your communication software and examine the Settings or Configuration entries relating to the modem, you likely will find a set of entries that allow you to specify different command strings. There are usually four different entries: the setup string, the send-before-dialing string, the send-before-answering string, and the send-after-hanging-up string.

The *setup string* is usually the most important command string for your modem, and is executed when the communication software starts. This is the command that lets you set specific operating characteristics for the modem. For example, the command ATZ gets the modem's attention (AT), then resets the modem to its power-up defaults. If there is room for a second setup string, additional commands can be entered to tailor the modem's overall operation.

The *send-before-dialing* string provides an opportunity to specify changes to the modem's dialing pattern (i.e., use pulse or TouchTone dialing) or modify dialing-related S-registers. Another handy use of the send-before-dialing string is to adapt the modem's dialing for unusual telephone setups. For example, you can disable call waiting by adding a *70 to the string; that way, the modem will automatically disable call waiting before it establishes a connection. It is also possible to introduce delays into the string, which allow the modem to wait for second (outside line) dial tones (typical in businesses using PBX equipment).

The *send-before-answering* string is rarely used in actual practice because a modem is rarely used to receive calls (unless you're running a BBS or attempting an unusual communications link). But in situations where the modem must answer a ringing telephone line, this string can be used to ensure a match with the calling modem's configuration.

Finally, the *send-after-hanging-up* string allows the modem to be reconfigured or reset after a communication link is terminated. In most cases, any entry in this string will be a reset command (i.e., ATZ), though it is also possible to use the send-before-dialing and send-after-hanging-up strings in concert to operate the modem in an unusual fashion, then restore the modem to a reset state afterward.

D

Testing Your Modem

OK, the new modem is installed, the communication software is loaded, the telephone line is connected . . . and nothing happens. This is an all-too-common theme for today's technicians and computer users. Although the actual failure rate among ordinary modems is quite small, it turns out that modems (and serial ports) are some of the most difficult and time-consuming devices to set up and configure properly. As a consequence, proper setup initially can simplify troubleshooting significantly. When a new modem fails to work properly, there are a number of conditions to explore:

- Incorrect hardware resources An internal modem must be set with a unique IRQ line and I/O port. If the assigned resources are also used by another serial device in the system (such as a mouse), the modem, the conflicting device (or perhaps both) will not function properly. Remove the modem and use a diagnostic to check available resources.

Reconfigure the internal modem to clear the conflict. External modems make use of existing COM ports.

- Defective telecommunication resources   All modems need access to a telephone line in order to establish connections with other modems. If the telephone jack is defective or hooked up improperly, the modem might work fine, but no connection is possible. Remove the telephone line cord from the modem and try the line cord on an ordinary telephone. When you lift the receiver, you should draw dial tone. Try dialing a local number; if the line rings, chances are good that the telephone line is working. Check the RJ11 jack on the modem. Bent connector pins can break the line even though the line cord is inserted properly.

- Improper cabling   An external modem must be connected to the PC serial port with a cable. Traditional serial cables were 25-pin assemblies. Later, 9-pin serial connectors and cables became common; out of those nine wires, only three are really vital. As a result, quite a few cable assemblies might be incorrect or otherwise specialized. Make sure that the serial cable between the PC and modem is a "straight-through" type cable. Also check that both ends of the cable are intact (i.e., installed evenly, no bent pins, and so on). Try a new cable if necessary.

- Improper power   External modems must receive power from batteries, or from an ac eliminator. Make sure that any batteries are fresh and installed completely. If an ac adapter is used, see that it is connected to the modem properly.

- Incorrect software settings   Both internal and external modems must be initialized with an AT ASCII command string before a connection is established. If these settings are absent or incorrect, the modem will not respond as expected (if at all). Check the communication software and make sure that the AT command strings are appropriate for the modem being used; different modems often require slightly different command strings.

- Suspect the modem itself   Modems are typically quite reliable in everyday use. If there are jumpers or DIP switches on the modem, check that each setting is placed correctly. Perhaps their most vulnerable point is the telephone interface, which is particularly susceptible to high-voltage spikes that might enter through the telephone line. If all else fails, try another modem.

CHECKING THE COMMAND PROCESSOR

The command processor is the controller that manages the modem's operation in the command mode. It is the command processor that interprets AT command strings. When the new modem installation fails to behave as it should, you should first check the modem command processor using the procedure outlined below. Before going too far with this, make sure you have the modem's user guide on hand if possible. When the command processor checks out, but the modem refuses to work under normal communication software operations, the software might be refusing to save settings such as COM port selection, speed, and character format.

1. Make sure the modem is installed properly and connected to the desired PC serial port. Of course, if the modem is internal, you will only need to worry about IRQ and I/O port settings.

2. Start the communication software and select a "direct connection" to establish a path from your keyboard to the modem. You will probably see a dialog box appear with a blinking cursor. If the modem is working and installed properly, you should now be able to send commands directly to the modem.

3. Type the command AT and then press the <ENTER> key. The modem should return an "OK" result code. When an "OK" is returned, chances are that the modem is working correctly. If double characters are displayed, try the command ATE0 to disable the command mode echo. If you do not see an "OK," try issuing an ATE1 command to enable the command mode echo. If there is still no response, commands are not reaching the modem, or the modem is defective. Check the connections between the modem and serial port. If the modem is internal, check that it is installed correctly and that all jumpers are placed properly.

4. Try resetting the modem with the ATZ command and the <ENTER> key. This should reset the modem. If the modem now responds with "OK," you might have to adjust the initialization command string in the communication software.

5. Try factory default settings by typing the command AT&F than pressing the <ENTER> key. This should restore the factory default values for each S-register. You also might try the command AT&Q0 and <ENTER> to deliberately place the modem into asynchronous mode. You should see "OK" responses to each attempt, which indicates the modem is responding as expected. It might be necessary to update the modem's initialization command string. If the modem still does not respond, the communication software might be incompatible, or the modem is defective.

CHECKING THE DIALER
AND TELEPHONE LINE

After you are confident that the modem's command processor is responding properly, you also can check the telephone interface by attempting a call; this also can verify an active telephone line. When the telephone interface checks out, but the modem refuses to work under normal communication software operations, the software might be refusing to save settings such as COM port selection, speed, and character format.

1. Make sure the modem is installed properly and connected to the desired PC serial port. Of course, if the modem is internal, you will only need to worry about IRQ and I/O port settings.

2. Start the communication software and select a "direct connection" to establish a path from your keyboard to the modem. You will probably see a dialog box appear with a blinking cursor. If the modem is working and installed properly, you now should be able to send commands directly to the modem.

3. Dial a number by using the DT (dial using tones) command followed by the full number being called; for example, ATDT15083667683 followed by the <ENTER> key. If your local telephone line only supports rotary dialing, use the modifier R after the D. If calling from a PBX, be sure to dial 9 or other outside-access codes. Listen for a dial tone, followed by the tone dialing beeps. You should also hear the destination phone ringing. When these occur, they ensure that your telephone interface dials correctly, and the local phone line is responding properly.

4. If there is no dial tone, check the phone line by dialing with an ordinary phone. Note that some PBX systems must be modified to produce at least 48 volts dc for the modem to work. If there is no dial tone, but the modem attempts to dial, the telephone interface is not grabbing the telephone line correctly (the dialer is working). If the modem draws dial tone, but no digits are generated, the dialer might be defective. In either case, try another modem.

UNDERSTANDING TYPICAL COM-MUNICATION PROBLEMS

Even when the modem hardware is working perfectly, the serial communication process is anything but flawless. Problems ranging from accidental loss of carrier to a catastrophic loss of data regularly plague computer communication. To make your on-line time as fool-proof as possible, this section shows you how to deal with some of the most pernicious communication problems.

Modem Settings

Modem settings are critically important to inter-modem communication. The number of data bits, use of parity, number of stop bits, and data transfer speed must be set precisely the same way on both modems. Otherwise, the valid data leaving one modem will be interpreted as complete "junk" at the receiving end. Normally, this should not happen when modems are set to auto-answer; negotiation should allow both modems to settle at the same parameters. The time that incompatible settings really become a factor is when negotiation is unsuccessful, or when communication is being established manually. A typical example is an avid BBS user with communication software set to 8 data bits, no parity bit, and 1 stop bit trying to use a network which runs at 7 data bits, even parity, and 2 stop bits. The aspect that stands out with incompatible settings is that virtually nothing is intelligible, and the connection is typically lost.

Line Noise

Where faulty settings can load the display with trash, even a properly established connection can loose integrity periodically. Remember that serial communication is made possible by an international network of switched telephone wiring. Each time you dial the same number, you typically get a different set of wiring. Faulty wiring at any point along the network, electrical storms, wet or snowy weather, and other natural or man-made disasters can interrupt the network momentarily, or cut communication entirely. Most of the time, brief interruptions can result in small patches of garbled text. This type of behavior is most prominent in real-time on-line sessions (such as typing in an e-mail message). When uploading or downloading files, file-transfer protocols can usually catch such anomalies and correct errors or request new data packets to overcome the errors. When you have trouble moving files or notice a high level of "junk" on-line, try calling back; when a new telephone line is established, the connection might be better.

Transmit and Receive Levels

Other factors that affect both leased and dial-up telephone lines are the transmit and receive levels. These settings determine the signal levels used by the modem in each direction. Some Hayes-compatible modems permit these levels to be adjusted. The range and availability of these adjustments is in large part controlled by the local telephone system. For example, the recommended settings and ranges are different for modems sold in the UK than for those sold in the US. See the documentation accompanying the modem to determine whether this capability is supported.

System Processor Limitations

Some multi-tasking operating systems occasionally can lose small amounts of data if the computer is heavily loaded and cannot allocate processing time to the communications task frequently enough. In this case, the data is corrupted by the host computer itself. This could also cause incomplete data transmission to the remote system. Host processor capabilities should be a concern when developing software for data communications when the line speed is greater than 9600 bps and the modem-to-DTE connection is 19200 bps or higher (for example, when data compression is used). The modem will provide exact transmission of the data it receives, but if the host PC cannot "keep up" with the modem because of other tasks or speed restrictions, precautions should be taken when writing software or when adding modems with extra high speed capabilities. One way to avoid the problem of data loss caused by the host PC is the use of an upgraded serial port such as a Hayes Enhanced Serial Port card or a newer modem with a 16550 UART. Such advanced modems are powerful enough to take some of the load off of the PC processor. When processor time is stretched to the limit, try shutting down any unnecessary applications to reduce load on the system.

Call Waiting

The call waiting feature now available on most dial-up lines momentarily interrupts a call. This interruption causes a click that informs voice call users that another call is coming through. While this technique is dynamite for voice communication, it is also quite effective at interrupting a modem's carrier signal, and it might cause some modems to drop the connection. One way around this is to set S-register S10 to a higher value so the modem tolerates a fairly long loss of carrier signal. Data loss might still occur, but the connection will not drop. Of course, the remote modem must be similarly configured. When originating the call, a special prefix can be issued as part of the dialing string to disable call waiting for the duration of the call. The exact procedure varies from area to area, so contact your local telephone system for details.

Automatic Timeout

Some Hayes-compatible modems offer an automatic timeout feature. Automatic timeout prevents an inactive connection from being maintained. This "watchdog" feature prevents undesired long-distance charges for a connection that was maintained for too long. This inactivity delay can be set or disabled with S-register S30.

System Lock Up

There are situations where host systems *do* lock up, but in many cases it is simply that one or the other of the computers has been *flowed off;* that is, the character that stops data transfer has been inadvertently sent. This can happen during error-control connections if the wrong kind of local flow control has been selected. In addition, the problem could be the result of incompatible EIA 232-D/ITU V.24 signaling. When systems seem to cease transmitting or receiving without warning, but do not disconnect, perform a thorough examination of flow control.

E

A s you explore on-line resources, you will encounter a myriad of different software. You've probably heard the terms *public domain, freeware, shareware,* and others like them. Your favorite bulletin board system (BBS) or FTP mirror site probably has many programs described by one or more of these words. There's a lot of confusion about and between these terms, but they actually have specific meanings and implications. Once you understand them you will have a much easier time navigating the maze of programs available to you, and understanding what your obligations are, or aren't, with each type of program. Let's start with some basic definitions.

PUBLIC DOMAIN

Public domain has a very specific legal meaning. It means that the creator of a work (in this case, the "work" is a piece of software) who had legal ownership of that work, has given up ownership and dedicated the work to "the public domain." Once something is in the public domain, anyone can use it in any way they choose. The original author has no control over use of the "work," and they cannot demand payment for its use.

To place a work in the public domain, the author must place a statement in the program code, documentation, or other materials explicitly stating the program's "public domain" nature. If you find a program that the author has explicitly put into the public domain, you are free to use it however you see fit, without paying for the right to use it (i.e., licensing). But use care; due to the confusion over the meaning of the words, programs are often described by others as being "public domain" when in fact they are shareware or free copyrighted software. To be sure a program is public domain, you should look for an explicit statement from the author to that effect. Because most people work for the promise of getting paid, true public domain software is generally quite rare.

COPYRIGHTED SOFTWARE

Copyrighted software is the opposite of public domain. A copyrighted program is one where the author has asserted their legal right to control the program's use and distribution by placing the legally required copyright notices in the program and its documentation. The law gives copyright owners broad rights to restrict how their work is distributed, and provides penalties for those who violate these restrictions. Virtually all commercial software is copyrighted. When you find a program that is copyrighted, you *must* use it in accordance with the copyright owner's restrictions on distribution and payment. Usually these are clearly stated in the program documentation and disk package.

Maintaining a copyright does not necessarily imply charging a fee, so it is perfectly possible and legal to have copyrighted programs that are distributed free of charge. Much of the software you find on-line (i.e., drivers and program patches) is copyrighted by the manufacturer, but it is available free to anyone who downloads them. Such programs are often referred to as *freeware,* though this term was in fact trademarked by the late Andrew Flugelman and the legality of its use by others could be questioned. In any case, the fact that a program is *free* does not mean that it is in the public domain, though this is a common confusion.

SHAREWARE

Shareware is copyrighted software that is distributed by authors through bulletin boards, on-line services, disk vendors, and copies passed among friends. It is commercial software that you are allowed to try out before you pay for it.

Shareware authors use a variety of licensing restrictions on their copyrighted works, but most authors who support their software require you to pay a "registration fee" (the purchase price of the software) if you continue to use the product after the trial period. Some authors indicate a specific trial period after which you must pay this fee; others leave the time period open and rely on you to judge when you have decided to use the program, and therefore should pay for it. Once you license the product, you get the full registered commercial version (and other varied benefits). Occasionally a shareware author requires registration but does not require payment; this is so-called "0$ shareware." Most companies provide copyrighted software, but you can find a wealth of shareware on commercial on-line services and bulletin boards.

The shareware system and the continued availability of quality shareware products depend on your willingness to register and pay for the shareware you *use*. The registration fees you pay allow authors to support and continue to develop their products. As a software user, you benefit from this system because you get to try the software and determine whether it meets your needs *before* you pay for it. Authors benefit because they are able to get their products into your hands with little or no expense for advertising and promotion. As a result, it is not unusual to find shareware products that rival retail software in quality and performance, but costs only a fraction of commercial software. Most reputable shareware authors are members of a trade association such as the *Association of Shareware Professionals* (ASP).

ASP members' shareware meets additional quality standards beyond ordinary shareware. Members' programs must be fully functional (not crippled, demonstration, or out of date versions); program documentation must be complete and must clearly state the registration fee and the benefits received when registering; members must provide free mail or telephone support for a minimum of three months after registration; and members must meet other guidelines that help to ensure that you as a user receive good value for your money and are dealt with professionally. The ASP also provides an Ombudsman program to assist in resolving disputes between authors and users.

For more information on the ASP or to contact the ASP Ombudsman, write to ASP, 545 Grover Road, Muskegon, MI USA 49442-9427, or fax 616-788-2765. You also can contact the Ombudsman on CompuServe via an electronic mail message to 70007,3536.

THE ASP OMBUDSMAN

TechNet BBS (owned and operated by Dynamic Learning Systems) is an approved BBS member of the ASP. The ASP wants to ensure that the shareware principle works for you. If you are unable to resolve a shareware-related problem with an ASP member by contacting the member directly, the ASP might be able to help. The ASP Ombudsman can help you resolve a problem or dispute with an ASP member, but it does not provide technical support for member's products or documentation. Please write to the ASP Ombudsman at 545 Grover Road, Muskegon, MI 49442-9427 USA, Fax 616-788-2765, or send a CompuServe mail message to the ASP Ombudsman at 70007,3536.

F

> # Electronics
> # Trade Associations

National Electronic Service Dealers Association (NESDA)
2708 W. Berry St.
Ft. Worth, TX 76109
T: 817-921-9061

Electronic Technicians Association International (ETAI)
602 N. Jackson St.
Greencastle, IN 46135
T: 317-653-4301
F: 317-653-8262

International Society of Certified Electronics Technicians (ISCET)
2707 W. Berry St.
Ft. Worth, TX 76109
T: 817-921-9101
F: 817-921-3741

Electronic Service Dealers Association (ESDA)
4927 W. Irving Park Rd.
Chicago, IL 60641
T: 312-282-9400

National Appliance Service Association (NASA)
9240 N. Meridian St., Ste 355
Indianapolis, IN 46260
T: 317-844-1602
F: 317-844-4745

North American Retail Dealers Association (NARDA)
10 E. 22nd St., Ste 310
Lombard, IL 60148
T: 708-953-8950
F: 708-953-8957

United Servicers Association (USA)
P.O. Box 626
Westmont, IL 60559
T: 708-968-6752
F: 708-719-0041

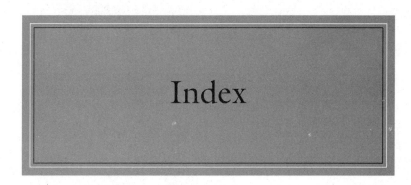

Index

Entries appearing in bold indicate artwork.

About the Author

Stephen Bigelow

Stephen J. Bigelow is an electronics engineer specializing in electronic design. He is the founder and president of Dynamic Learning Systems, a technical research and writing firm with a special focus on PC and peripheral service topics. He has written several technical books, including *Troubleshooting and Repairing PC Drives and Memory Systems, Troubleshooting and Repairing Computer Monitors, Second Edition,* and *Troubleshooting, Maintaining and Repairing Personal Computers: A Technician's Guide.* The author has also written and published nearly one hundred technology-related feature articles in industry magazines such as *Popular Electronics NOW,* and *Computer Craft.* His company is based in Marlboro, Massachusetts.

Tired of fixing your PC "in the dark"?

Now you don't have to! Subscribe to:

The **PC Toolbox**™

Hints, Tips, and Fixes for Every PC User

A publication of Dynamic Learning Systems, P.O. Box 282, Jefferson, MA 01522-0282
Tel: 508-829-6744 Fax: 508-829-6819 BBS: 508-829-6706 Internet: sbigelow@cerfnet.com

Finally, there is a newsletter that brings computer technology into perspective for PC users and electronics enthusiasts. Each issue of *The PC Toolbox*™ is packed with valuable information that you can use to keep your system running better and longer:

♦ Step-by-step PC troubleshooting information that is easy to read and inexpensive to follow. Troubleshooting articles cover all aspects of the PC and its peripherals including monitors, modems, and printers.

♦ Learn how to upgrade and configure your PC with hard drives, video boards, memory, OverDrive™ microprocessors. scanners, and more.

♦ Learn to understand the latest terminology and computer concepts. Shop for new and used systems with confidence.

♦ Improve system performance by optimizing your MS-DOS® and Windows™ operating systems.

♦ Learn about the latest PC-oriented web sites, ftp locations, and other online resources. Check out our web site at http://www.dlspubs.com/

♦ Get access to over 2000 PC utilities and diagnostic shareware programs through the Dynamic Learning Systems BBS.

There is no advertising, no product comparisons, and no hype - just practical tips and techniques cover to cover that you can put to work right away. If you use PCs or repair them, try a subscription to *The PC Toolbox*™. If you are not completely satisfied within 90 days, just cancel your subscription and receive a full refund - *no questions asked!* Use the accompanying order form, or contact:

Dynamic Learning Systems, P.O. Box 282, Jefferson, MA 01522-0282 USA.

MS-DOS and Windows are trademarks of Microsoft Corporation. The PC Toolbox is a trademark of Dynamic Learning Systems.
The PC Toolbox logo is Copyright©1995 Corel Corporation. Used under license.

366

The PC Toolbox™

Use this form when ordering *The PC Toolbox™*. You may tear out or photocopy this order form.

YES! I'm tired of fixing computers in the dark! Please accept my order as shown below: (check any one)

_____ Please start my *one year subscription* (6 issues) for $39 (USD)

_____ Please start my *two year subscription* (12 issues) for $69 (USD)

PRINT YOUR MAILING INFORMATION HERE:

Name: Company:

Address:

City, State, Zip:

Country:

Telephone: (   )   Fax: (   )

PLACING YOUR ORDER:

By FAX: Fax this completed order form (24 hrs/day, 7 days/week) to 508-829-6819

By Phone: Phone in your order (Mon-Fri; 9am-4pm EST) to 508-829-6744

By Web: Complete the online subscription form at http://www.dlspubs.com/

___ MasterCard   Card: ___ ___ ___ ___ ___ ___ ___ ___ ___ ___ ___ ___ ___ ___ ___ ___

___ VISA   Exp: ___/___   Sig: _____

Or by Mail: Mail this completed form, along with your check, money order, PO, or credit card info to:

Dynamic Learning Systems, P.O. Box 282, Jefferson, MA 01522-0282 USA

Make check payable to Dynamic Learning Systems. Please allow 2-4 weeks for order processing. Returned checks are subject to a $15 charge. There is a 90 day unconditional money-back guarantee on your subscription.

Have we missed your site?

This book is dedicated to being a first-rate reference for electronics enthusiasts, technicians, and engineers. If you know of any web sites, BBS numbers, or any other online resources that we have not included in this edition, please let us know so that we can include it in the next edition. Contributors will receive a free issue of *The PC Toolbox*™*.

TELL US ABOUT THE SITE(S)

Site name:

Resource(s):
(i.e. URL, BBS, FTP, etc.)

List any other sites here:

TELL US ABOUT YOURSELF

For taking the time and trouble to help us, we would like to thank you by sending you a complementary copy of *The PC Toolbox*™. If you would like a copy, please fill out your mailing information below:

Name:

Address:

City: State: Zip:

Country:

GETTING THE INFORMATION TO US

Fax this form to us at: **508-829-6819**

Send us an E-mail at: **sbigelow@cerfnet.com**

Mail this form to us at: ***Dynamic Learning Systems***, PO Box 282, Jefferson, MA 01522-0282

368